SYBIL CAVANAGH grew up in Fife and is a gradu
Diploma in Librarianship at the University of Wa
Glasgow, where a long-standing interest in Scottish history developed into a particular interest in local history. She came to West Lothian in 1984 and has been Local History Librarian with West Lothian Libraries since 1990. She is the author of several modest local history booklets, and a contributor to a new history of Bathgate.

KNEALE JOHNSON started his career as a petroleum engineer with BP, working in the Middle East and Canada, and was involved in the early days of North Sea activity. After six years as a consultant he rejoined BP and took charge of UK onshore exploration and production in the early 1980s, including planning the expansion of the Wytch Farm oilfield in a highly sensitive area in Dorset. Since 1990 he has again worked as a self-employed consultant. His experience of managing environmentally sensitive projects has been invaluable in his work for BP dealing with the legacies of the Scottish shale oil industry.

JOHN H. MCKAY came to Pumpherston in 1933 at the age of four, where his father became foreman plumber at Pumpherston Oil Works. After leaving school he worked for a short time at the Oil Works before National Service. He then served for 33 years in Customs and Excise. He graduated from the Open University in 1975 and later gained a PhD for a thesis on *The social history of the Scottish shale oil industry*. In 1974 he was elected to Edinburgh District Council, and from 1984 to 1988 was Lord Provost of Edinburgh. Since then he has enjoyed retirement but has also been a part-time tutor with the Open University.

JAMES O'HAGAN has lived in Pumpherston for more than 60 years. Student vacations were spent working with Scottish Oils both as painter and laboratory technician. He taught at the local High School for 20 years, spent a few years as greenkeeper on the golf-course and passed another decade writing a weekly outdoors column for *The Scotsman*. He performed with some mediocrity for most of the village teams and later tried to make amends by serving as secretary to the Golf Club, the Junior Football Club, a Youth Club and the Gala Day Committee.

Pumpherston

The Story of a Shale Oil Village

JOHN H. McKAY
SYBIL CAVANAGH
JAMES O'HAGAN
KNEALE JOHNSON

Edited by
SYBIL CAVANAGH

Luath Press Limited
EDINBURGH
www.luath.co.uk

First Edition 2002

A copy of a disc giving detailed references and sources for Parts One and Two of this book is available at the Local History Library, West Lothian Library HQ, Hopefield Road, Blackburn, EH47 7HZ. Tel. 01506 776331.

The paper used in this book is acid-free, neutral-sized and recyclable.
It is made from low chlorine pulps produced in a low energy, low emission manner from renewable forests.

Printed and bound by
J. W. Arrowsmith Ltd., Bristol

Typeset in 10.5 point Sabon by
S. Fairgrieve, Edinburgh, 0131 658 1763

Acknowledgements

The assistance of the following people is gratefully acknowledged.

Mr William Armstrong
Mr Brian Cavanagh
Mr John Clark
Mrs Catherine L. Denholm
Mr Alan Docharty
Mrs Betty Dungate
Mrs Lena Findlay
Mr William Gold
Mrs Elizabeth Henderson
Mrs Elizabeth Hepburn
Mr W.T. Johnston
Mr George Kerr, Robertson Partnership
Mrs Marjorie Lamond
Mr Ian Loch
Mrs Moira McCreight
Mr Jim McGinty
Mr Colin MacKerracher
Mrs Jean MacLardy
Miss Ann W. Maxwell
Mrs A.A. Mulholland
Mr Hislop Murphy
Rev. John Povey
Miss Helen Scott
Mrs Vicky Whyte
Mr Joe Wood
and many others too numerous to mention, who contributed photographs, memories and information.

Mrs W. Armstrong, for permission to extract information from the history of Pumpherston written by her late husband, Mr Vic Armstrong.
Dr Robin Chesters, Almond Valley Heritage Centre, for permission to use photographs from the collection.
The Ordnance Survey, for permission to reproduce OS maps.
Miss J.D. Slater for drawing the early maps of Pumpherston estate, based upon estate plans in the National Archives of Scotland, used by permission of the Marquess of Linlithgow.
The staff of the Midlothian Local Studies Library, West Lothian Local History Library, BP Grangemouth and the BP Archives, University of Warwick.
Templecrone Co-operative Society, for permission to reprint extracts from *My Story* by Paddy the Cope.
BP Amoco and West Lothian Council, for financial assistance.

Contents

Introduction

A VILLAGE WITH A population of under 1,400 would hardly seem to merit a history of this size or scope, but then Pumpherston is no ordinary village.

Its history is closely bound up with that of the Scottish shale oil industry, the first commercial oil industry in the world, and so the book looks first of all at that pioneering Scottish industry. The history of the Pumpherston Oil Company is seen in the context of the wider industry, in which from 1914 it was the dominant company.

Pumpherston village grew up rapidly from the mid 1880s, and is a classic example of a company village. The oil companies provided work and housing; they also provided various recreational facilities and assisted many more, so that an active social, cultural and sporting life evolved. The paternalism of the companies is strange to modern eyes, but was accepted almost without thought at the time.

Just over a century after being set up, the shale oil industry closed down – the victim not of any failure by the industry to innovate, but of foreign competition. BP has recently completed the rehabilitation of the Oil Works site, and a new eighteen-hole golf course in a green landscape has emerged from the grime and pollution of the old Works. BP commissioned this study of Pumpherston to mark the end of the long association between the village and the oil industry.

Foreword

ONE CONSEQUENCE OF WORKING for a big company is that you find yourself doing a good many things that just 'go with the job'. That's how I ended up sitting on a good many committees, well-intentioned bodies who'd sometimes lost track of their original purpose. Being chairman of Young's Paraffin Light and Mineral Oil Company in a way 'went with the job' too – but the experience could hardly have been more different, more meaningful or more memorable. The overwhelming sense of history and the chance to do voluntarily today what clearly feels so right for tomorrow was simply unique.

Over a century ago the developing shale industry led to the growth of the village of Pumpherston. By the 1920s the existence of a workforce skilled in the handling of petroleum was crucial in the decision to site the next phase of the oil industry at nearby Grangemouth. Some 50 years later Grangemouth in turn became a vital link in bringing ashore the newly discovered North Sea oil.

Until relatively recent times the Works at Pumpherston was the core of the community, not only providing employment but also a framework for the social fabric of the village. By the time of final closure in the early 1990s the village was no longer reliant on a single industry, but the legacy the Works would leave future generations was still in the balance. Many industries whose time has passed leave scars, or worse, on the landscape long after they disappear. I believe that what has been achieved at Pumpherston is to change that legacy, leaving instead an inheritance in the shape of a golf course that can benefit the whole community. To me that's a really fitting way to end the industrial phase of the site's remarkable life.

I wholeheartedly welcome this book which so vividly brings together the whole story of an industry, a village and its community.

John Williams
Chairman
Young's Paraffin Light and Mineral Oil Company Ltd
July 2001

PART ONE

Pumpherston and the Shale Oil Industry

by

JOHN H. McKAY

General view of Pumpherston Oil Works looking west, c.1950. Left and centre, the coking stills with their drum-like tops. The shallow circular dishes (centre) are 'stell bottoms' for the stills.
(BP archives)

The Development of the Shale Oil Industry

THE VILLAGE OF PUMPHERSTON owes its existence to the formation of the Pumpherston Oil Company Limited in 1883 and, until well after the end of the Second World War, the life of the village centred on the Oil Works. The history of the village must include the story of the Company and the history of the Company can only be understood in the context of the shale oil industry as a whole.

The history of the industry falls fairly naturally into four phases:

- 1851-mid 1870s – from the setting up of Young's works at Bathgate until adverse economic conditions puts the small oil companies out of business.
- mid 1870s-c.1895 – the formation of a number of limited companies bringing with them a substantial capital investment.
- c.1895-1918 – expansion continues through the exploitation of fresh shale fields by existing concerns rather than the creation of new enterprises.
- 1919-1962 – from the amalgamation of the shale oil companies into Scottish Oils Limited, a subsidiary of the Anglo-Persian Oil Co Ltd (later Anglo-Iranian and eventually British Petroleum), until the closing down of the whole shale oil industry in 1962 (although the Pumpherston Works continued to produce detergent and refine wax until the early 1990s, using imported oil).

CHAPTER ONE

The Early Years of the Shale Industry, 1851-77

THE SCOTTISH OIL INDUSTRY WAS FOUNDED by James Young (later known as 'Paraffin' Young), a Glasgow chemist working in Manchester for Charles Tennant and Company. In 1848 Young, in partnership with fellow chemist Edward Meldrum, began production of lubricating oils from a seepage of crude oil found in a coal mine at Alfreton in Derbyshire. This source of raw material was soon exhausted. Because it had been found in a coal seam, Young believed that crude oil could be derived from coal and he experimented with samples of coals from all over Britain. Through a friend, Hugh Bartholomew, the manager of a Glasgow gas works, it was found that cannel coal from Boghead, near Bathgate in West Lothian was a suitable material. By the end of 1850 a new works was under construction at Bathgate for E.W. Binney & Co. This firm was composed of Young, Meldrum and Edward Binney, a Manchester lawyer who had assisted in the establishment of the Alfreton enterprise, and who provided much of the capital for the Bathgate works.

In the first ten years of its existence the firm enjoyed a virtual monopoly in the British oil trade, largely because Young had carefully drawn up patent specifications and was willing to go to law to enforce them. Over little more than ten years, production increased tenfold, and profits rose from £5,000 a year in 1851 to £57,000 in 1864. All three partners made their fortunes.

By 1864 the raw material, cannel coal, was becoming scarce. There was also some competition from the exploitation of oil shale, the distillation of which was not covered by Young's patent. Young's patent, however, did cover the refining of crude oil and it was not until its expiry in 1864 that the way was completely open for the expansion of the fledgling shale oil industry in the Almond valley, particularly around Broxburn, Oakbank and West Calder. These developments from 1863 to 1866 have been described as the Scottish Oil Mania. In all a total of 43 works can be identified in the Almond valley and 44 different firms or individuals were concerned in their operation. In 1861 the Valuation Rolls indicate only one oil works in the area, the Bathgate plant of E.W. Binney & Co. In 1863 there were nine, in 1864 eleven, in 1865, twenty and in 1866, 27. The number of works operating at the same time peaked at 30 in 1868 and declined gradually until in 1877 only seven works remained. This was largely a consequence of generally bad trade conditions along with competition from American oil producers. Imports of

American oil almost trebled from 1873 to 1878. One or two oil works survived in other regions of Scotland, notably in Renfrewshire. A small works at Straiton near Loanhead also continued in operation.

Many of the 43 concerns in the Almond valley had operated in a very small way. However the seven surviving works belonged to four limited companies, all of which had some influence on the further history of the industry. The most important of these was Young's Paraffin Light and Mineral Oil Company Ltd established in 1866. James Young had in 1864 bought out his two partners in E.W. Binney & Co. He then began to build a large works at Addiewell, opening mines at West Calder to supply it and the Bathgate plant with shale. These mines, the Bathgate works and the partially completed plant at Addiewell were transferred to the new limited company in 1866 and Young was paid £403,000. He remained a director, but had little to do with the day-to-day running of the firm and left the industry completely in 1870. Although beset by problems arising from over-capitalisation, Young's Paraffin Light and Mineral Oil Company Ltd dominated the industry until the late 1870s and remained an important element thereafter.

The other three concerns which survived the bad conditions of the mid 1870s all had some connection with James Young and the first oil works at Bathgate. On the dissolution of the firm E.W. Binney & Co., Young's partner, Edward Meldrum, joined forces with George Simpson, a local coalmaster and entrepreneur, and Peter McLagan, owner of the Pumpherston estate and land at Stankards, to establish works and mines at Boghall and Uphall Station. This Uphall Mineral Oil Company Ltd built a crude oil works near Winchburgh. At Oakbank, near Mid Calder, mines and works were set up by a partnership made up of Sir James Young Simpson of chloroform fame, S.B. Hare of Calder Hall, the mineral landlord, and William McKinlay, formerly manager of the company mining cannel coal at Boghead for E.W. Binney & Co. In 1869 the works and mines were taken over by the Oakbank Oil Company Ltd under the management of Norman Henderson who had been an engineer and draughtsman with E.W. Binney & Co. The last was the Dalmeny Oil Company Ltd whose works and mines near South Queensferry were managed by James Jones who had been a foreman engineer with the Bathgate company.

From 1877 the history of the industry is that of companies fighting for survival against often intense foreign competition. There was also a fierce rivalry among the Scottish companies themselves. It is therefore appropriate at this point to describe briefly the industry, its methods and its equipment as they were in 1877, the end of the period of consolidation.

Mining methods and oil production until 1877

MINING METHODS

The production of oil was in three stages. First the cannel coal or shale had to be mined; it was then distilled in a dry state to produce crude oil; this was then refined by distillation and treatment with chemicals. The refined products included burning oil (which we now know as paraffin or kerosene), wax, lubricating oils, light oils for the paint and rubber industries and gas oil, so called because of its use in the enrichment of coal gas.

Mining was by the normal methods of the time and was governed by the geological conditions. The oil shales form part of the Carboniferous Limestone series of rocks which lie under the coal measures of central Scotland. The Oil Shale group, which occurs only in the Lothians and Fife, forms the upper part of this series and there are seven main seams, which lie within a vertical distance of some 3,000 feet. Since being laid down, the sedimentary strata of which they are part have been subjected to major geological disturbance resulting in much faulting and folding. There are four major faults and innumerable minor ones. As a result the seams are very rarely level but lie at angles. In the Drumshoreland basin, between Broxburn and Pumpherston, some are almost vertical. This had both favourable and unfavourable consequences. There were many outcrops of shale on the surface. It was often possible to extract the mineral by opencast working or by shallow drift mines. As the industry expanded, larger quantities of shale were required and the disadvantages became apparent. The numerous faults made it difficult to gain access to large areas of shale from a single shaft or mine.

In 1873, when shale mines were first included in the Annual Reports of the Inspectors of Mines, it is clear that mines had moved on a great deal from the early opencasts, but they remained relatively small. In seventeen shale mines in the Lothians and Fife in 1875, there were 977 miners, an average of 57 per mine. The largest employed 124 men and only three employed more than 100. The smallest had only twenty workmen.

Until 1877 the 'longwall' method was well suited to the shale seams which were about three or four feet in thickness. 'Longwall' involved removing the shale in one operation along the length of a long face. The face moved forward into the seam as the shale was removed. Access to the face was by roadways, twelve or thirteen yards apart and about seven feet wide. As the face advanced the roadways were formed and protected by the construction of wooden pillars some three and a half feet square, at intervals of three or four feet on each side of the roadway. These were filled with any waste material and 'brushings' from the roof of the roadway. As the face moved forward the roof was allowed to come down to meet the pavement in the 'cundy' or empty space left by the removal of the shale.

The seam was worked by 'holing' into the material immediately below the shale. This was done by the pick if the material was soft or by explosives if not. In the early days, the 'holing' was into a layer of fireclay or clod soft enough to be removed by hand. This was removed to a depth of about three feet. While doing this the miner was protected by 'sprags' or short lengths of prop placed at an angle to support the shale. When holing was completed along the length of the face the 'sprags' were removed and the shale brought down by explosives. It was then 'backened' or thrown back to the roadhead to be shovelled into hutches. During these operations the roof was supported by two rows of props close to the working face. As the face advanced, the row farthest from the face was moved forward in a leap-frog movement and their place in supporting the roof was taken by extending forward the rows of pillars forming the sides of the roadways.

The shale was brought to the surface in 'hutches' or small wagons each bearing a 'pin' or marker indicating the miner who had produced the shale in it. The hutches were brought from the working face to the 'lye' at the pit bottom, or, in inclined mines, at the lower end of the main 'dook'. The lye was simply a siding in the mine railways where full or empty hutches could be left to await transport to the surface or the working area. The hutches were often pushed by manual labour, but in some mines horses were used. From the 'lye' the full hutches were taken to the surface by winding in cages up vertical shafts, or were hauled in 'rakes' or on carriages up inclines.

OIL MANUFACTURE

Mining methods were highly labour-intensive, and the same is true of the early processes in oil manufacture. The shale was often carted long distances to the retorts. On reaching the retorts the shale had first to be broken down to a suitable size, often described as about the size of a man's fist. With larger pieces the heat could not penetrate evenly through the shale. If the pieces were too small, however, they had a tendency to fuse in the retort. Initially, shale breaking was done manually. However by the mid 1870s, the larger firms had installed mechanical shale breakers.

The shale had then to be heated to produce vapours that would condense into crude oil. Shale does not contain oil as such. It contains various compounds of carbon and hydrogen, together with some nitrogen and sulphur, resulting from the decomposition of vegetable matter at the time the sedimentary rocks were laid down. The shale was heated to some 700 degrees Fahrenheit in retorts. At first these were modified coal gas retorts, consisting of a cast-iron tube with a door at one end and a pipe at the other for the removal of the vapours. The retorts were set horizontally in brick ovens heated by coal fires. After charging, the door was sealed or 'luted' with clay and screwed up tightly. The volatile products were driven off over a period of between sixteen and twenty-four hours. The spent shale was removed through the end door and the retort freshly charged.

The throughput of these retorts was only some six or seven hundredweights (300 to 350 kilograms) per day, and by the mid 1870s most of them had been replaced by vertical retorts. As the name implies these were made up of cast iron tubes set vertically in brick ovens. A charging hopper at the top was controlled by a ball valve and the bottom was set in a shallow tray of water to prevent the escape of gas. They operated continuously. At intervals a small amount of spent shale was raked out through the water 'lute' which served to quench the shale. At the same time a corresponding quantity of fresh shale was introduced into the retort from the charging hopper. The retort was thus kept full of shale. The gases were removed by a pipe at the top. The vertical retorts had a capacity of from 25 to 30 hundredweights (1,250 to 1,500 kilograms) a day and were the most significant advance in technology in the period up to 1877.

The vapours from the retorts were passed through atmospheric condensers to produce a mixture of oil and ammonia water which was allowed to separate in tanks, the oil rising to the top and being run off. At some works the ammonia water was allowed to run to waste and was a source of considerable pollution in the local streams. In works where the shale was richer in ammonia, it was worthwhile converting the liquor into sulphate of ammonia by evaporation. This was

Part of the sulphate of ammonia plant at Pumpherston Refinery. The ammonia was crystallised in the drum-shaped evaporators at the top, then put in the 'spin dryers' at the bottom of the picture to dry the crystals.
(BP)

The paraffin wax sheds. The paraffin released from the pipes (top) filled up the trays, then dripped down from tray to tray, cooling as it went and forming stalactites of wax.
(Almond Valley Heritage Centre)

a valuable nitrogenous fertiliser and was always a source of profit to the industry. There remained a large volume of 'permanent' gases which could not be condensed into liquid hydrocarbons. In the early stages of the industry this was allowed to escape to the atmosphere after 'scrubbing' in towers filled with coke to remove any residual liquid hydrocarbons.

The crude oil was refined by distillation and treatment with chemicals. The product of well run horizontal retorts could often be treated without prior distillation. The crude oil from vertical retorts required distillation in cast-iron pot stills before the chemical treatment. The oil was then mixed with between five and ten per cent of its volume of sulphuric or hydrochloric acid. The acid rose to the top of the mixture bringing impurities with it in the form of a layer of acid tar which was removed. In some works this was run to waste, polluting local streams. At others the mixture was utilised in the production of sulphate of ammonia. After the acid treatment, the crude oil was mixed with an alkali to neutralise any residual traces of acid.

The saleable products were obtained by distillation in horizontal cylindrical stills of malleable iron, which, because of their shape were known as boiler stills. They were heated by coal fires or later by steam. As the lighter oils were taken off, more crude oil was added to maintain their production. After a period, heavier oils accumulated in the still and when these reached a certain level, the crude oil feed was stopped and the heavy oil was distilled off, leaving a residue which was distilled in pot stills, producing grease and coke.

The saleable end products of the whole process were naphtha for the paint and rubber industries; burning oil (paraffin) for illumination; and lubricating oil which gave rise to one of the important products of the later stages of the industry. Paraffin wax affected the lubricating qualities of oil and had to be removed. In the early days it was thought to be useless and was disposed of, but the mechanisation of candle making in the mid 1850s provided a market for it, and wax became an important part of the industry's products.

The wax was removed by cooling which made it crystallise; this was followed by draining through hemp bags, pressure filtration, washing in naphtha and treatment with charcoal. Up to 1877 the only significant improvement to this process was the invention of a compression refrigerator for cooling the wax, designed by Alexander Kirk, the engineer and manager at Bathgate Chemical Works.

THE PARAFFIN SHEDS

'The Paraffin Sheds were cauld to work in – at the coolers, there were long pipes, great long pipes covered in snaw. As a lad, I had to gather a bucket of snaw frae the pipes, twice a day, for the chemists tae pit samples in it to cool. When you first went to work there, the auld hands they'd say, see if there's no a smell doun there, and you'd bend down to sniff at the pipe, and they'd open the release valve and it was sulphate of ammonia in your face – you couldnae breathe.'

At the end of its first phase the industry was one of relatively simple technology, cheap enough to allow many small firms to set up. By 1877 all of these had ceased operation, leaving only four limited companies operating seven relatively substantial works. It is clear that at this stage the dominant company was Young's Paraffin Light and Mineral Oil Company Ltd with a paid-up capital of nearly half a million pounds. In 1876 it produced nearly four and a half million gallons of oil. The Uphall Mineral Oil Company Ltd had been reorganised in 1877 as the Uphall Oil Company Ltd with a paid-up capital reduced from £200,000 to £170,000. It operated works at Uphall Station, Benhar near Fauldhouse, and at the Hopetoun Works near Winchburgh, and overall was rather less than half the size of Young's company. The other two survivors were on an even smaller scale. The Oakbank Oil Company Ltd had a paid-up capital of £45,000 and had works capable of producing one million gallons of crude oil and refining one and a half million gallons. The smallest, the Dalmeny Oil Company Limited, had a paid-up capital of £18,900 and produced crude oil only.

CHAPTER TWO

Recovery, Consolidation and Expansion, 1877-87

TRADING CONDITIONS IMPROVED BETWEEN 1876 and 1877. The value of the products of a ton of shale at Oakbank works rose by over 20%. Production increased and shale output went up by 27%. The number of miners employed by the surviving companies rose from 644 to 751. Profits and dividends recovered: that of Young's company almost doubled and the Oakbank company payment trebled.

As well as increased activity by the surviving concerns, the improved trading conditions tempted the firm of Thomas and James Thornton to resume operations at Levenseat near Fauldhouse and at Hermand near West Calder. This concern was incorporated as the Hermand Oil Company Ltd in 1885.

The prosperity also encouraged new firms to start up. The first and best known was the Broxburn Oil Company Limited, established in 1877 to take over the many small works around Broxburn which had all come into the hands of Robert Bell, the mineral lessee who had been the first in Scotland to distil oil from shale. At the same time Bell sold the lease to the new company. The driving force behind this was William Kennedy who had been general manager of the Oakbank company. The new company discarded the old horizontal and vertical retorts and constructed a new crude oil works to the north of Broxburn, based on a retort designed by Norman Henderson who had been the works manager at Oakbank. For fuel it used the permanent gases produced by the retort itself and the retort was filled and emptied directly from hutches. The 'Henderson' retort provided a significant saving in fuel and labour costs and was a major advance in technology. The cost of producing crude oil was reduced by almost 30 per cent. The retort was said to produce a more uniform product, the quality of which was not, as in the older retorts, entirely dependent on the skill of the operatives. The oil was also easier to refine and gave a higher yield of paraffin scale.

The Broxburn company was the success story of the industry's second phase. From 1880 to 1886, good dividends of 25% were paid. In 1885 the paid-up capital was almost £200,000, and the works were using 1,000 tons of shale per day to produce ten million gallons of oil per year. This kind of success encouraged competition. The established companies found that their profits fell as those of the Broxburn company rose. Young's company's dividend fell from 17.5% in 1877 to only 4% in 1883, and in 1882 and 1883 the Oakbank and Uphall companies paid no dividend at all. The obvious solution was to invest in the new retorts which had proved so successful at Broxburn and in fact the Uphall and Young's companies began to do so in 1880. Oakbank held back, preferring to experiment with improvements to existing vertical retorts.

The adoption of the new retort coincided with the introduction of new mining methods, more suited to thicker seams of shale. From its start the Broxburn Oil Company worked the five foot seam of Broxburn shale by the 'stoop and room' system. This is best described as driving a mine or a 'dook' from the pit bottom or from the surface into the shale until it met the edge of the seam, generally defined by a geological fault line. At intervals along the dook, 'levels' were driven out to the edge of the seam. From these, 'upsets' were driven to connect each level with the next. Thus a system of criss-crossing tunnels was constructed throughout the whole of the accessible seam. The tunnels were the 'rooms' and the pillars of shale left between them were the 'stoops'. When this 'first working' was complete, the stoops were removed, starting at the remotest part of the seam and working back to the dook head. At both stages the full depth of the seam was taken out, so the method was not suitable for seams which had layers of unproductive material or 'fakes'. The stoop and room method was far more productive than the other method. In the 1880s, mines worked thus produced an average of over 550 tons per man-year, compared with around 350 tons in those with longwall faces.

The success of the Broxburn company encouraged the formation of a number of new companies. In 1878, the Straiton Oil Company (taken over in 1882 by the Midlothian Oil Company) began operations at Loanhead. Also in 1878, the Binnend Oil Company (in 1881 taken over by the Burntisland Oil Company) started in Fife. In 1880 the British Oil and Candle Company (taken over by the Lanark Oil Company in 1883) reopened mines and works at Tarbrax and Lanark. In 1883 the Westfield Oil Company started up in Fife, the West Lothian Oil Company restarted the defunct Boghall works, and the Bathgate Oil Company commenced business at Seafield. 1883 also saw the formation of the Pumpherston Oil Company. In 1884 the Holmes Oil Company was formed to produce crude oil at a small works between Broxburn and Uphall. Also in 1884 the Linlithgow Oil Company established mines, crude oil works and a refinery at Ochiltree, Champfleurie and Kingscavil near Linlithgow. All of these were limited liability companies; the exception was the private partnership of James Ross & Company which in 1885 opened mines and crude oil works at Philpstoun near Linlithgow.

There were also a number of firms working on 'coalfield shales' in Lanarkshire and Ayrshire. These were of relatively little importance in the industry; most did not survive into the 1890s and the last closed down completely in 1903. From the 1880s onwards the industry was virtually confined to the Lothians, Fife and Renfrew. After the demise of the Walkinshaw and Burntisland companies and the transfer of the Clippens company to Straiton, the shale oil industry existed entirely in the Lothians. On the closure of the Straiton mines and works of the Clippens company it was confined solely to the valley of the River Almond.

To operate a shale oil company at this stage needed a substantial capital investment. The 21 limited companies active at one point or other during the period from 1877 to 1887 had something over one million pounds invested in the industry. In present day values this might amount to as much as 200 million pounds.

During the period 1877-87, improvements were made to the design of retorts. Sulphate of ammonia had become an increasingly valuable product as its price rose from about £14 per ton in 1869 to £21 in 1882. The new Henderson retort, although its designer claimed that it gave a higher yield of sulphate of ammonia than the older verticals, was considered capable of improvement. The problem was that the temperature which produced the best quality of crude oil was very much lower than the temperature which produced the maximum quantity of ammonia. Experiments were carried out in the late 1870s and early 1880s with retorts operating at two temperatures. A number of new designs were patented, the most successful being that by William Young of the Clippens Oil Company Ltd, and by George Beilby of the Oakbank company.

The Young and Beilby patent retort had a top section of cast iron twelve feet long, in which the shale was heated to the temperature required for oil production. This was mounted on a ten foot section made of firebrick which could withstand the high temperatures needed for ammonia. The retorts were heated by their own permanent gases, supplemented by coal gas from a producer retort set in the same bench. The main source of trouble was the joint between the iron and firebrick sections, which had a tendency to leak when allowed to cool. This problem was overcome by continuous operation of the retorts, thus bringing to an end the weekend shut-down. The new retort became available in 1883 and it quickly superseded the Henderson patent in the successful companies.

This period was not one of unalloyed success. Of the fifteen companies set up between 1877 and 1885, only seven continued into the 1890s; and of these the Burntisland Oil Company Ltd failed in 1892, and the Linlithgow, Holmes and Hermand companies struggled on into the first years of the twentieth century. The companies which survived to make a contribution to the third and most successful phase of the industry were, from the first phase, Young's, Dalmeny and Oakbank; and from the second phase, Broxburn, Pumpherston and James Ross & Company.

One factor which contributed to success or failure was *when* the new technology was adopted. Two of the older companies, Young's and Uphall, had begun to install the Henderson retort in 1880. Two of the new companies, Burntisland and Linlithgow, were (in 1882 and 1884 respectively) committed to the Henderson retort at a time when the Young and Beilby retort was being developed. The

Straiton and West Lothian companies invested in a modified Young and Beilby retort made entirely of firebrick which proved to be unreliable.

However, the adoption of the Young and Beilby retort did not of itself guarantee success. The Lanark company installed two benches but because of management problems and the high price paid for the acquisition of the Whitelees and Tarbrax works, this company failed after only two years.

CHAPTER THREE

Pumpherston Oil Company and the Industry, 1883-1914

Formation

THE PUMPHERSTON COMPANY EVENTUALLY became the largest concern in the shale oil industry, so it is worth examining its progress to see why it was so successful when so many other companies failed.

The prime mover in the formation of the Company was William Fraser, who until 1883 had been manager of the Uphall Oil Company. Since 1877 the Uphall company had been a relatively poor performer, paying dividends totalling less than 20% over a period in which the Broxburn company paid nearly 120%. However it had access to good shale fields, which attracted Young's company and resulted in a merger with the Uphall Company in 1884.

In 1883 Fraser left the Uphall Oil Company and with his brother Archibald leased 1,200 acres of shale fields at Pumpherston and Houstoun from the landlords, Peter McLagan and Colonel Shairp. The Pumpherston Oil Company Limited was incorporated on the 3 November 1883. The nominal capital (what the subscribers, the people asking for the formation of the company, could seek as capital) was £130,000 in 13,000 £10 shares. A prospectus inviting subscriptions for 7,000 shares was issued with a closing date for applications of the 30 October 1883.

Applications were received for 26,418 shares. In the end the 7,000 shares were allocated to 205 shareholders. The 190 men and fifteen women were drawn from a fairly wide spectrum of Scottish society. Most were in the more prosperous sections of the population. 31 were in commerce – merchants, stockbrokers, bankers etc; 35 were professionals such as doctors, engineers, teachers, clergymen, etc; eighteen were engaged in manufacturing or production as ironfounders, coalmasters, builders, etc; sixteen were in retail trade as grocers, drapers, etc. In addition, there were a chief constable, a farmer, a land steward, a publisher and a sub-editor of the *Glasgow Herald*. Men in managerial positions were also well represented; a sugar house manager; a railway inspector; colliery managers; oil company managers, etc. There were also coal salesmen, cashiers, commercial travellers in less senior employed positions. In what might be termed the working classes were clerks, colliery oversmen, an office porter, engine drivers, a locomotive fireman and a packer.

101 of the shareholders lived in Glasgow and the west of Scotland; 25 in Edinburgh and district and 64 in the immediate area of the Almond valley of West Lothian. There were a few shareholders further away including five in the Borders, one in Fort William and one in Birmingham.

William Fraser and Archibald Fraser had been allocated shares to the value of £15,000 as payment for the shale leases. (These were known as vendors' shares.) William Fraser was the largest shareholder. The smallest shareholding was of five shares (£50 nominal) and the average was £439 nominal. More than a quarter of the shareholders held shares with a nominal value of £100 or less.

The shareholders had made an initial payment of 10s on application and a further £1 on allotment. At intervals during 1884 further calls totalling £7 (£1 on the vendors' shares) were made. The paid-up value of each share (i.e. what the shareholder actually paid for each share) was £8.10s. During this period therefore, the Company had £61,500 available for the construction of works, sinking mines and building houses for the workmen.

The Workforce

ORIGINS OF THE WORKFORCE

Fourteen hundred people arrived in Pumpherston between 1884 and 1891. A significant proportion of them were Irish. Driven by poverty and attracted by the hope of a better life, the Irish emigrated in their millions. Emigration to Scotland was cheap and easy, and reached a peak in the 1890s.

By the time of the 1891 Census, the Pumpherston Oil Works had been established, and the village of Pumpherston had been built. The total population of Pumpherston in 1891 (including Pumpherston Mains and Cottages) was 1,389. Of these, 265 were Irish born (19%). If to these are added the Scottish-born children of Irish parents, the total is 425 (30%) or nearly one in three.

Pumpherston's 19% Irish born was far higher than the figure for West Lothian as a whole, which in 1891 was 6.8%. Part of the reason may be the huge number of jobs created in the 1880s as the Oil Works were set up and the mines sunk. The immigrants tended to settle in concentrations of Irish, helped by chain migration; the earlier immigrants invited their relatives and friends over, gave them a bed and found them a job. In Pumpherston in 1891, of nearly 200 lodgers and boarders, over half were Irish. Indeed, 42% of all the Irish-born in Pumpherston in 1891 were lodgers and boarders, and a high proportion of them lived with Irish families. Almost without exception these lodgers were young – in their twenties and unmarried. Some certainly returned to Ireland; others would later have married and settled in Scotland.

Figures for Irish-born residents of Pumpherston are not yet available after 1891, but it is presumed that Irish immigration dropped away rapidly after the

A Gala Day procession through Pumpherston before the First World War. On the left are some of the Auld Raws, the first housing built in the village.
(Hislop Murphy)

First World War as it did in the rest of Scotland and in West Lothian. The Irish integrated into the rest of the population, especially after the testing times of the First World War and the bringing of Catholic schools into the state system in 1918; and that integration has become even more effective as mixed marriages have become more common in the last 20 or 30 years.

19% of Pumpherston's new residents came from Ireland; what of the other 81%? Over a third of them were born in the shale area – West Lothian and those parts of Midlothian lying to the west of the Pentlands. Another 123 (9%) were born in Midlothian, east of the Pentlands, including Edinburgh. Only thirteen were native to Mid Calder, suggesting that there was little movement by local workers into the new industry. Some 22% were born in Stirlingshire, and 17% in Lanarkshire, including Glasgow. Twenty-one were born in England, and four overseas.

All this suggests that many men who already had experience working in other shale mines and oil works came to Pumpherston. As a relative latecomer to the industry, the Pumpherston company could take advantage of this fact, by recruiting experienced workers from the earlier firms. This was especially important in the skilled posts such as cooper, engine fitter and boilermaker. The foreman paraffin refiner at Pumpherston in 1891 had been working in the industry since 1859 and came to Pumpherston with a wealth of experience in the oil trade. It seems clear that the Pumpherston company was able to recruit capable, experienced men for responsible positions in the new Works and this must have contributed greatly to the Company's success.

Another striking feature of Pumpherston in 1891, as of most new industrial villages of the period, was the gender imbalance. 831 (60%) of the population

PADDY THE COPE

Of all the Irish who came to West Lothian, only one set down his story in writing – Paddy the Cope. Patrick Gallagher came to Scotland about 1890 to seek seasonal work at the harvests, then returned for several years to work as a shale miner in the Uphall and Winchburgh areas.

'We bought the usual steerage tickets, but did not go to the steerage that night. Sally and I sat up on the deck all night. When we landed in Glasgow we got the train for Uphall and went to Mrs McMahal's house in the Randy Rows. There was one bed in the kitchen, two in the room. There was no scullery. She had eight lodgers but she was glad for us. Before she went to Scotland, she was a neighbour of Sally's, and she insisted on our staying with her until we got a house of our own. Sally and the landlady slept in the kitchen. Four of the lodgers were on the night shift and four on the day. We stayed there for a week. The second day after arriving, I got work in the Holmes mines.'

were male, and only 558 (40%) were female. This was an inevitable result of the high number of young, male unmarried workers who lodged or boarded. Also, the sons of the house could get jobs in the mines or Works; most of the daughters had to leave home to find work. No women were employed in the Works although a small number were listed as working in domestic service or in shops. One was described as a nurse. Only two married women had occupations, a tailoress and a laundress.

Occupations

In 1892, when entertaining a party of visitors from the British Association, Archibald Fraser, the secretary and general manager of the Company, stated that 700 people were employed by the Pumpherston Oil Company. The census of 1891 lists 483 men and boys in the village in employment, 469 of whom can be firmly identified as employed in the mines and Works in 1891. The other 231 employees of the Company were men commuting daily from places outside the village, such as Mid Calder or Uphall.

Not surprisingly the workforce was relatively young. Only 71 men were aged over 40. The youngest employee was an eleven year old messenger. Six boys aged thirteen, all sons of other workmen, were listed as employed, three as miners, one as a retortman, one as a brickmaker and one as a bricklayer's labourer. As many as 132 were aged twenty or under. The oldest man was 68. Indeed, a striking aspect of the new village of Pumpherston was its youthfulness. The oldest person in Pumpherston was Agnes Hill, aged 71, a widow keeping house for her shale-

General View of the Oil Works looking west, 1940s/50s. Left, the coking stills. Far left in the background, the wax extraction plant. Tiered tower, right of centre, the cracking plant. The railway line was the Camps Branch line of the North British Railway, leaving the main line at Camps Junction near Kirknewton, crossing the Almond by the Camps Viaduct, and rejoining the Bathgate-Edinburgh line near Uphall Station.
(BP)

miner son at 108 Pumpherston Rows. A surprising 38% of the population of Pumpherston were under the age of fourteen. Today the proportion under fourteen is nearer 22%.

Of the 483 men and boys identified as working in the Oil Works and elsewhere, 205 were heads of households, mainly married men, although a few were widowers. 190 were lodgers or boarders; 67 the sons of other workers. Twenty workers were related in other ways: brothers, brothers-in-law, fathers, etc.; and one was described as a visitor. The high number of lodgers and boarders is characteristic of the shale communities at this time.

Some 237 of the workers were shale miners. Of the rest, three were managers – the Works manager, the mine manager and his under manager. Only five could be described as professionals: an apprentice mining engineer; a chemist's assistant; a lithographic artist; a student for the civil service and a licentiate of the United Presbyterian Church.

There were at least eight foremen: two foremen engineers; a foreman oil refiner; foreman paraffin refiner; foreman labourer; retorts foreman; foreman (sulphate of ammonia department); and foreman bricklayer. There were 48 skilled tradesmen of various kinds: blacksmiths, bricklayers, plumbers, engineers, joiners, coopers, clerks, etc. There were 62 men with skills acquired otherwise than by apprenticeship. These included retortmen, stillmen, a locomotive engine driver, engine keepers, etc. Of the men and boys, 108 were in unskilled jobs such as labourers, stokers, boiler firemen etc.

The Management

The first directors were chosen from the subscribers – the people requesting the formation of the limited company. The first was James Wood, described as a coalmaster of Bathville, by Armadale. Wood was in fact a substantial figure in Scottish commercial life. He was born in 1840, the son of a Paisley handloom weaver, and started working life as a clerk to the Portland Coal Company. In 1860 he set up on his own account as a retail coal merchant, and soon established a wholesale business with depots around Glasgow. In 1871 he acquired mineral leases at Armadale in West Lothian. He bought the estate of Bathville and lived there, and by 1880 he was a major figure in the Scottish coal industry with collieries at Bathville, Barbauchlaw and Hopetoun in West Lothian and at Shields and Shieldsmuir at Motherwell. He acquired brick works, founded the Bathville steel works and was the majority shareholder in the Armadale Iron Company Ltd. In 1903 he bought the estate of Wallhouse near Torphichen and was elected to West Lothian County Council. Henceforth he lived mainly at Wallhouse, where he died in 1933, aged 93, leaving considerable sums to various charities.

Another subscriber was James Craig, an engineer with a Paisley firm which was a major manufacturer of textile machinery, but which, as the oil industry developed, branched out into the construction and supply of oil-related equipment.

Other subscribers were Robert Brown Tennent, an ironfounder at Whifflets near Coatbridge; and John Paterson, a partner in the firm of Romanes and Paterson, tartan and tweed drapers, Princes Street, Edinburgh.

Also a subscriber was Archibald Fraser, who had been book-keeper with

THE DIRECTORS

William Fraser was a moving spirit in the shale oil trade. Shotts-born, he was a young mining manager in 1881, living with his wife and two daughters in Uphall. By 1883 he was the manager of the Uphall Oil Company Ltd. Although only thirty years old, he already had some experience of management in both the mining and oil works sides of the industry. He was appointed managing director of the new Pumpherston company. The other directors were James Wood, James Craig, Robert B. Tennent and John Paterson. The commercial and industrial experience of these men, combined with Fraser's practical knowledge of the industry, were to be important in the Company's development. As secretary, Archibald Fraser, his brother, brought his experience of a major industrial concern.

Robert Addie & Sons, ironmasters, of Bothwell and Glasgow. Apart from William Fraser himself, the last subscriber was John Milligan Fraser, almost certainly another Fraser relative. (William Fraser's mother was Isabella Milligan, and Milligan was his own middle name.) John Milligan Fraser was described as a coal salesman of 40 St Enoch Square, Glasgow, the address of James Wood's Glasgow office.

The auditor was Thomson McLintock, CA, who had founded what was to become one of the major accounting firms of twentieth century British business.

Reasons for success or failure

The other successful companies had subscribers and directors drawn from the same spectrum of Scottish commerce, men with a great deal of experience and success in various aspects of Scottish business at the highest level.

The companies which failed in the second period were generally lacking in men of this experience. For example, the Hermand Oil Company had as its first directors a coalmaster turned farmer, a local banker and the previous owner of the mines and works. They had experience of the practical side of the industry but none of the wider commercial world in which the industry was beginning to operate. The Holmes Oil Company was promoted by a group made up largely of retail shopkeepers in Wishaw.

Another factor was that many of the unsuccessful firms – the Binnend, Westfield, Lanark and West Lothian companies – were promoted to purchase mines and works which were grossly overvalued. Only the West Lothian Company lasted more than a few years and even it failed after eight years of unprofitable working.

Some of the companies experienced management problems. From 1851 to 1866, the largest concern in the industry was under the personal management of James Young. One of the achievements of John Orr Ewing, first chairman of Young's Paraffin Light and Mineral Oil Company Ltd, was to oversee the transition from this personal management to 'that of the directors and their managers'. This was a feature of the industry from that time and managers became an important element in the success of any company – and in some of the failures. The Linlithgow Oil Company laboured under a number of disabilities including a shale field which did not come up to expectations. However, in 1892 it was disclosed by the directors that oil vapour from the retorts had been allowed to escape along with furnace fumes. This and other inefficiencies were attributed to faults in the late manager who among other failings had not 'got on very well with the men'. The Lanark Company board reported that 'the management at the crude oil works and the refinery did not work harmoniously'.

These management problems were probably inevitable. In the period from

Ammonia compressors and refrigerating plant used in the paraffin extraction plant at Pumpherston Refinery. It seems unlikely that the floral arrangement was a permanent feature.
(BP)

1877 to 1887, the industry in the Almond Valley developed from four companies using over half a million tons of shale, to one of a dozen companies with a capacity four times that. The established companies, of course, had had the first choice of capable managers – for example, Norman Henderson at Oakbank then Broxburn; George Beilby at Oakbank; and James Jones at Dalmeny.

The newer companies were therefore at some disadvantage which was aggravated by technological changes during this period. This was particularly noticeable in the retorting of shale. In 1877 the industry, which had survived the poor conditions of the mid 1870s, was one of relatively simple technology. The introduction of the considerably more complex Henderson retort and particularly the Young and Beilby version, meant alterations to working practices, and difficulties and expense for the management. Also important was the changeover from the personal management of individuals like Young, to management by boards of directors assisted by salaried managers. This was probably the area which gave rise to the most problems, at least for the Pumpherston company.

CHAPTER FOUR

Pumpherston Oil Company – The Early Years

THE FIRST MEETING OF THE Board of Directors of the Pumpherston company was held on 6 November 1883. James Wood was appointed chairman for three years. The main item of business was the appointment of a Works manager. The post was offered to A.C. Thomson, the son of a farming family near West Calder. He was thirty years old and had been assistant manager at the works of the Walkinshaw Oil Company Ltd near Paisley. New equipment had recently been installed there, so with this experience Thomson must have seemed the ideal candidate.

It is interesting that the early meetings of the Board were held mainly in William Fraser's home at Uphall and later at the Works' office at Pumpherston, although the registered office of the Company was in Glasgow. This may indicate that the directors were very much concerned with the detail of the construction. Whatever the reasons, it is clear that meetings at Pumpherston gave the directors close control over the day to day running of the Company.

There is indeed some evidence in the minutes of these early meetings to indicate that the Board would exercise a tight control over the Company's affairs. On 8 January 1884, the chairman and Mr Craig were instructed to call upon the Company's law agents to ask for a reduction in their bill for £158.14s.4d – their expenses for setting up the Company. The bill was reduced to £140. The same meeting approved a draft of the Company's business arrangements submitted by the secretary, and conferred upon the managing director power 'to do everything requisite for the successful prosecution of the business of the Company'.

The directors immediately set about the building of the Works. Tenders were sought for freezing machines, winding engine, hutches, tanks, filter presses, steam boilers, workshop machinery, still tops, oil washers, roofing, still bottoms and oil boilers, all plant and equipment usually found in a shale oil works. The managing director and the Works manager were left to enquire into the retort to be adopted. At the first Annual General Meeting of the Company on 29 February 1884 it was reported that the Company hoped to be making and refining oil by the month of July or August in time for the winter demand.

By 21 April 1884 it had been decided that three benches of Young & Beilby retorts would be erected. By the end of May 1884, £24,196 had been spent on plant and two mines were being sunk. Also in May the first fatal accident took place in the Company's mines, when John McCormack was killed by runaway hutches. His widow claimed £250 from the Company but settled for £135.

So confident of success were the directors that it was decided to expand the refinery plant to deal with the output of crude oil from the Holmes company – the product of two benches of Young & Beilby retorts – thus increasing the refinery throughput by two thirds. However on 9 October a special meeting of the directors was held to consider delays in starting the Works, caused partly by an accident at the Works, but also by deficiencies in the operation of ammonia stills installed by the manager. The stills had to be replaced. By the end of October burning oil and sulphate of ammonia had to be bought from other companies to fulfil orders and this was still necessary in November. On 27 October, Mr Thomson, the Works manager, resigned feeling that his services were not acceptable to the board of directors. (He later went on to become chief constructional engineer with Scottish Oils.) William Fraser, the managing director was to manage the Works until a replacement was found. During this period the Company appointed agents in London, Manchester, Belfast and Dundee.

At the end of March 1885, Mr G.D.H. Mitchell of the Oakbank company was appointed manager at a salary £100 per annum greater than his predecessor. This may indicate the difficulties experienced in attracting a suitable person. Another management appointment in August 1884 was significant because it brought to Pumpherston what became one of its best known families. Mr James Caldwell, underground manager at Benhar for Robert Addie & Sons was appointed mining manager.

Despite all the difficulties, the first four months of working (ending on 30 April 1885) resulted in enough profit to pay a dividend of 10% on the paid up capital as well as providing for depreciation. In June it was agreed to erect a fourth bench of retorts. In August the secretary was authorised to visit the 'chief

JAMES CALDWELL

Like William Fraser, James Caldwell began work in a coal mine when he was just a child. He had little formal education, but he was in charge of the village library and, by his own exertions, he qualified as a mining manger. He was employed by companies at Morningside, Arden and Benhar before moving to the Pumpherston Oil Company in 1884.

Working with William Fraser, he extended the Company by finding four workable seams of shale and expanding production at Pumpherston and Seafield. He was involved with the introduction of the new Bryson retort and additional shale fields at Deans, Tarbrax, Breich, Whitehill and Blackburn.

He was described as being able to 'express himself forcibly' when occasion demanded but was known to give credit where it was due and was generally regarded with respect by colleagues and workers. According to newspaper reports he retired in late 1920 which means he was still working at 80 years of age. He took little part in public life but he was a J.P. and a member of Mid Calder School Board. He died in August 1921 aged 81, leaving four sons and three daughters.

towns on the continent' and appoint agents with a view to opening up a continental trade. By the end of November agencies had been established in Rouen, Paris, Lille, Berlin, Magdeburg, Leipzig, Vienna, Frankfurt and Cologne.

By the end of 1885, after a little over a year in operation, the Company began to experience severe financial problems. The scale of the Works actually built was considerably greater than what had been contemplated at the formation of the Company, and in fact some £104,000 of capital expenditure had been incurred, over a period when only £61,500 of capital had been subscribed by shareholders. There ensued a period of constant juggling of bank overdrafts, debenture bond issues, loans from banks and private individuals, and it is clear that chairman James Wood, managing director William Fraser and other directors made substantial loans to the Company at this time as well as acting as guarantors for loans from other sources.

Management problems

Also towards the end of 1885, there appeared the first indications that all was not well with the management of the Works. Alterations to the sulphate of ammonia plant were postponed because it was said that Mr Mitchell, the manager, was too busy with other extensions in progress. In March 1886 he made a complaint of interference with his authority. In April complaints were received from almost every customer regarding the colour and quality of the Company's semi-refined wax. This was due to the failure of the manager to complete the wax refinery department, almost four months after the scheduled starting date. The managing director claimed that the Works generally were being mismanaged, most seriously in the case of the paraffin refinery department. It was practically impossible to put matters right because his instructions were being ignored. In an attempt to improve matters, the board appointed an assistant manager, but warned the manager that in future he was to make himself responsible to the managing director.

Despite these problems the Company's first full year of operation ended on 30 April 1886 with enough profit to pay a dividend of 10%. At the Annual General Meeting it was decided to issue a further 4,000 shares to ease the difficulties in the capital account. A fifth bench of Young & Beilby retorts was to be erected.

Management problems continued through the summer and autumn of 1886, culminating in a board meeting which discussed a proposal by the chairman to resolve matters by means of a new agreement with the manager. The managing director, William Fraser, strongly disagreed with this course of action, saying that he had repeatedly urged the board to dispense with the manager's services on the best terms possible. The Company had suffered severely through the board's indecision, while the manager, Mr Mitchell, had brought the Company to the verge of ruin. Despite this the board decided to give the manager another chance.

William Fraser objected so strongly to this decision that he resigned from the board.

There was no improvement. Instructions from head office were ignored by the manager and there were continuing complaints about product quality. Matters came to a head in February 1887 when the manager forcibly ejected the cashier from the Works. He also sacked one of the clerks, 'using violence to the boy'. Despite letters from the secretary and the chairman, the situation remained unchanged, with the manager keeping a man as sentry at the office door. This farcical situation brought the board to its senses and William Fraser was persuaded to resume his seat. Mitchell was sacked and William Fraser's brother Archibald was appointed general manager in addition to his post as secretary. In accepting the position, he insisted that he should have complete control of every department and that there would be no interference from the directors. He regarded as crucial to the success of the Company the appointment of William Fraser as consulting director.

Management re-organisation

On the advice of William Young one of the designers of the Young & Beilby retort, the management of the Works was reorganised. He advised that, with the increasing complexity of the industry, it was impossible for one man to have full knowledge of all aspects of the work. Each department would require an experienced man at its head, responsible to an overall manager. After some initial disagreement, the board appointed as Works manager James Bryson, an engineer and under manager with Bairds of Gartsherrie. This was possibly one of the most important appointments made by the Pumpherston company. James Bryson

JAMES BRYSON

James Bryson was born in Coatbridge in 1850, and like his management predecessors was largely self-taught having had to leave school at 13 after the death of his father in a mining accident. He was an engineer with Summerlee Iron Company, Scott's Shipbuilding, Carron Iron Company and Bairds before arriving in 1887 as Works manager at Pumpherston Oil Company. With William Fraser he expanded the workings to Seafield, Livingston Station and Tarbrax.

He is credited with significantly reducing the cost of producing shale oil by the introduction of the superbly efficient Bryson or Pumpherston retort. In 1910 he was made a director of Pumpherston Oil Company and in 1915 became joint managing director. He was general Works manager when the shale oil companies amalgamated under Scottish Oils Ltd in 1919.

He was also active in public life, being a J.P. for Midlothian, a member of Mid Calder School Board, vice-chairman of the Midlothian Health Insurance Committee, preses of Bridgend Church and president of the Midlothian Liberal Association. He died at his home Ballengeich House in January 1930, leaving a widow and two daughters.

remained with the Company for the rest of his life, eventually becoming joint managing director and he played a very significant role in its success.

The result of the year's working in 1887 produced no dividend – probably due to the management problems. However, this was also a period of general difficulty in the trade due to foreign competition and over-production at home and in fact only the Broxburn, Dalmeny, Burntisland and Holmes companies declared dividends, and these were much reduced. The unfavourable trade conditions persuaded the oil companies to attempt to reduce wages, in both mines and Works. This was resisted by the men, particularly the miners and in July 1887 a strike began, mainly but not exclusively in the Broxburn area. It lasted 21 weeks.

The Pumpherston company, in its particularly delicate financial position, was anxious to keep working and offered a compromise of half the proposed reduction. When this was refused by the men, the Company recruited workmen in Bathgate and brought them to the Works by special trains. It is reported that over 300 pickets met the first train and that a large body of police was in attendance at Pumpherston Works. A miner was arrested for throwing stones at the train. The Company's efforts were frustrated by strikers picketing at Bathgate and persuading the 'blacknebs' to cease working. The Company repeated its offer of work at half the proposed reduction; this was accepted and work resumed on 1 August. The Company suffered the displeasure of the Scottish Mineral Oil Association but the Pumpherston mines and Works operated normally until the end of the dispute, which was in fact settled on similar terms.

James Bryson, managing director and general Works manager at Pumpherston.
(Almond Valley Heritage Centre)

In 1888 only the Oakbank, Broxburn, Hermand and Pumpherston companies paid dividends. These were small but the industry recovered and experienced a short period of relative prosperity. This was partly due to an agreement with the American oil companies regarding a major product of the industry, wax scale, the raw material for the production of candles. This agreement regulated the price and also the quantities to be produced by the various companies. The net result was that the amount distributed to shareholders rose from £19,000 in 1888 to over £90,000 in 1889, but fell back again to £57,000 in 1892.

The Pumpherston company shared in this general prosperity and declared dividends of 10% on its ordinary share capital each year from 1889 to 1892. The Company's difficulties with its capital structure had been partially resolved by the issue of additional shares raising another £70,000. 1891 also saw the end of a long-running dispute with the Holmes company. When starting up, the Pumpherston company had contracted for three years to purchase the crude oil produced from two benches of retorts at Holmes. The contract had been the subject of constant dispute about the required quantities not being delivered and about the method of calculating the price. Arbitrators were involved and the case went to the Court of Session and eventually the House of Lords where it was decided in the Pumpherston company's favour. In fact, the settlement was made on the same basis as an offer by the Company in the first year of the contract.

Foreign Competition

In August 1892, a party of members of the British Association visited Pumpherston Oil Works, and the report in the *West Lothian Courier* gives a description of the Works at that time. The daily output of shale was about 600 tons and the shale was taken straight to the hoppers feeding the shale breaker and from there to the retorts – 424 of them, in six benches. The ammonia liquor and crude oil were then condensed and separated. The refinery was in close proximity to the retorts and was capable of dealing with 120,000 gallons of crude oil per week. Archibald Fraser, the secretary and general manager explained that the Works covered 23 acres and employed 700 workers. The capacity of the Works when originally built was intended to be 250 tons of shale per day, but the plans had allowed for an extension which had indeed taken place. He claimed that the layout was such that there was an 'almost continuous series of operations in which the handling of materials is reduced to a minimum, and thus an enormous saving is effected in time and working expenses'.

Mr Fraser also referred to the severe difficulties being experienced by the shale oil companies at the time. The American companies operated at a considerable advantage in that their crude oil gushed out of the ground, and was got at virtually no cost compared with the processing necessary for shale oil. He referred to what he described as a war going on between the American and Russian trades, but declared that although the 'Scotch' trade might be harassed and curtailed for a time, it could not be completely extinguished. The Scottish oil industry was not free to fix its own prices: it must 'accept the prices fixed by the competing natural product, and then... conduct its own operations in such a way as to leave a margin of profit, if possible'. The seriousness of this competition can be seen in the Scottish industry's declining share of the UK market. In 1887, the shale oil production was some 52 million gallons and represented about 40% of UK consumption. In 1913, the Scottish industry produced 72 million gallons, but this was only 14% of the market. Imports of oil from Russia, the US and other countries rose to a total of nearly 600 million gallons just before the First World War.

The effects of this competition were made worse by the breakdown of the wax scale producers' agreement in 1892. The companies which had candle-making plants – Young's, Broxburn and Linlithgow – enjoyed an advantage over those – principally Pumpherston and Oakbank – which depended on the sale of the raw material to independent candlemakers. The Pumpherston company withdrew from the Association following a dispute about over-supply.

American competition and the flooding of the market by huge supplies of wax from the failed Burntisland Oil Company resulted in a catastrophic fall in the price of wax. For the year ending April 1893, Young's company alone reported a fall of £72,000 in the value of sales. Although the American agreement was revived for a few months in 1895, the Scottish industry from 1892 onwards had to compete in an unrestricted market with no artificial restraints on prices or production. In 1893 the distributed profit of the industry fell to just under £8,000; and in 1894 no dividends were paid on the ordinary shares of any of the companies.

CHAPTER FIVE

The Bryson Retort and Technical Refinements

THE OIL COMPANIES NATURALLY tried to compensate for the fall in the price of their products by reducing costs. Prices for materials such as coal and acid were beyond their control, but appeals were made, sometimes successfully, to the mineral landlords for a reduction in the royalties that the companies paid to the landowners for each ton of shale they mined. Some of the hardest hit companies were able to persuade their workforce to accept a cut in wages.

What was needed however was to find some way of cutting costs that was not dependent on outside factors. As in previous times of difficulty, this was found in further improvements to the industry's basic equipment, the retort. A number of different improved versions were designed and patented, and one of these, at Broxburn, was successful enough to let the Company build the Roman Camp crude oil works.

But the major improvement in retort design originated in the Pumpherston company where the Works manager, James Bryson, undertook a radical re-examination of retort technology. Until then it was thought that retorts had to be of small diameter to ensure that none of the shale was far from the source of heat. The lateral dimensions had not changed much since the first vertical retorts. The larger capacities of the Henderson and the Young & Beilby retorts were achieved by increases in length and the shale spent a longer time in them. There was no corresponding increase in throughput. Even the improved retort at Roman Camp works processed rather less than half as much again as the older versions. Bryson designed a continuously operating retort where the column of shale was kept moving downwards by frequently drawing out small quantities of spent shale. This movement made the shale move horizontally within the retort, and the very slow progress of the shale through the retort helped the penetration of heat to all parts. The final version of his design had a capacity of 150 cubic feet and a throughput of five tons of shale per day, almost three times as much as any of its predecessors. It was filled directly from hutches and unloaded directly into more hutches, all by gravity, so it brought savings in labour costs. Fuel costs were reduced,

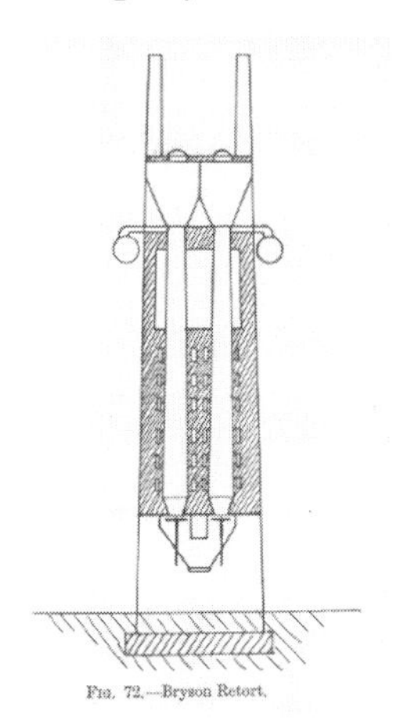

The Bryson (or Pumpherston) Retort, whose improved design contributed significantly to the success of Pumpherston Oil Company.
(West Lothian Council Libraries)

because after the initial firing, it was heated almost entirely by its own permanent gases. One bench of the new retorts replaced two benches of Young & Beilby retorts with a saving in labour of 32 men. This retort was the basis of one of the longest periods of consistent success experienced by the industry.

The Bryson retort created a feeling of some confidence in the industry, at least among the companies able to finance the new equipment. The chairman of the Oakbank company reported to his shareholders in 1903 that shale oil could now be produced at a cost approaching that of natural petroleum.

However, the cost of a bench of the new retorts was some £13,000, more than double that of the Young & Beilby version and this caused the demise of the smaller, less successful companies. The directors of the Linlithgow company in 1900 reported to the shareholders that the heavy outlay for modern retorts had been a serious difficulty. The Company closed in 1902, followed by the Holmes company, the Hermand company and the Caledonian company at Tarbrax.

The larger more solidly based companies could afford to invest in the new retorts. Young's company replaced the 1,024 Young & Beilby retorts in its three works, but decided not to expand capacity in case the additional production was difficult to sell in a bad year. Other companies were less cautious. The Broxburn company had already built a new crude oil works at Roman Camp. The Oakbank company replaced all its old retorts and started to develop a new shale field at Duddingston near Winchburgh, constructing the Niddry Castle crude oil works to deal with the shale. At the same time the Company arranged to take the entire crude oil production of the Dalmeny company and eventually took over the Dalmeny company in 1915.

The company which benefited most from the new technology was of course the Pumpherston company itself. Two benches of the new retort were installed in the defunct Bathgate Oil Company's works at Seafield which had been acquired in 1891. A further two were built at the Deans works of the West Lothian Oil Company, acquired in 1896, and the Works at Pumpherston was re-equipped with three benches. Altogether in 1899 seven benches of the new retort had a capacity of some 400,000 tons of shale per year, an increase of two-thirds in the Company's operations. In 1890 the Company had employed 336 miners underground, less than 8% of the total of 4,267 in the thirteen companies then active. In size, it was ranked fourth behind Young's company, Broxburn and Clippens. By 1899 the Pumpherston company's 554 underground workers made up 16% of the 3,374 men employed by the ten oil firms, and the Company was now in third place behind Young's and Broxburn.

In 1904 William Fraser acquired the shale fields and oil works at Tarbrax, near West Calder. The Tarbrax Oil Company Limited was formed to take these over. The works had three benches of retorts but another was added in 1909. The capital was £135,000 and although the company existed as an independent concern, the board of directors was the same as that of the Pumpherston company and all the crude oil produced at Tarbrax was refined at Pumpherston. In 1913 the Tarbrax company was completely absorbed by the Pumpherston company.

In 1914, after the amalgamation with the Tarbrax Oil Company Ltd, the Pumpherston company had 1,336 underground workers employed in mines at Pumpherston, Deans, Seafield, Mid Breich, Tarbrax and Cobbinshaw – almost a third of the 4,200 men employed in the six surviving companies. Pumpherston had become the largest shale oil company, and was also of considerable importance in the Scottish economy as a whole. The shareholders felt the benefit of the Company's success; in the years from 1900 to 1914, dividends totalling £728,053 were paid.

The workforce participated to a lesser extent in this prosperity. In the period before the Great War, the average wage in the oil works had increased by some 35% since the mid 1880s; in the shale mines the average was some 23% higher. This was in a period when the cost of living was relatively stable, so the workforce was substantially better off than it was when the Pumpherston company was set up.

Though paying such lavish dividends to its shareholders, the directors did not neglect to maintain and improve the mines and Works. During the same period, over £300,000 was set aside for depreciation, and capital expenditure of over £250,000 was incurred entirely from the internal resources of the Company. Nearly £100,000 was placed in a retort renewal fund and £128,000 was transferred to the reserve fund. The other shale companies enjoyed similar success during this period but none were able to match the Pumpherston company's performance.

In 1914, at the beginning of the Great War, the Pumpherston Oil Company Limited was the leader in an industry which was to be of considerable importance to the country during that conflict.

CHAPTER SIX

The Great War, 1914-18

IN THE EARLY MONTHS of the Great War, the Pumpherston company, and the shale industry generally, experienced severe problems. At the Annual General Meeting of Pumpherston shareholders in June 1915, a dividend of only 10% was declared on the ordinary shares, a very poor result compared with the previous year's 25%. However, this was considerably better than the other companies, none of which had any dividend for their ordinary shareholders. It was the War which was causing the downturn. Two of the principal products at Pumpherston were wax and sulphate of ammonia, which were mostly exported. During the early months of the Great War there was considerable disruption of exports, partly due to enemy action but also to the government's control of the export of wax and sulphate of ammonia. This restriction reduced the Company's profits. It was also more difficult in wartime to get supplies of materials, such as coal, pit wood and chemicals.

However, the chairman was able to report to the shareholders that, in the last few months of the financial year, these difficulties had largely been overcome by raising their prices and making special efforts to secure supplies. There was still however a shortage of labour. In May 1916, the chairman of the Pumpherston company reported to the Annual General Meeting of the shareholders that 537 (or a fifth) of the Company's employees had joined the armed forces. It had not been possible to fill all the vacancies. That the Works had been kept going at virtually full capacity was due to the men 'having maintained commendable regularity in their attendance at work'. A few women had been taken on at the mine heads and in the Works, '...but the nature of the operations generally was not favourable to the employment of female labour to any great extent.' However, enough women were employed for the trade unions to take an interest. In November 1916, the Oil Workers' Union asked for an increase for the women employed in the Oil Works, while at the same time the Shale Miners' Union requested a similar increase for the women employed at the mine heads. In February 1917, a Special Arbitration Tribunal awarded women aged eighteen and over an increase of 2s 6d per week bringing their wages to 22s 6d for a week of 54 hours.

Shortage of labour was a continuing problem. In April 1917, it was reported that, although some men had been transferred from Tarbrax, part of the third and the whole of the fourth bench of retorts were idle. The situation became particularly acute when conscription was introduced in 1916. The Director of Navy

Contracts wrote to the Pumpherston company in November 1916, stating that 'in view of the importance of the present output of shale oil being maintained it is very desirable that the company should take the necessary steps to obtain protection for its workmen.' All men at the Works and mines became exempt from military service.

WILLIAM FRASER

In June 1915 there occurred the death of William Fraser at the age of 62. Apart from a few months in 1886-1887, he had been managing director of the Pumpherston company from the start. Under his stewardship the Company had become the leading firm in the shale oil industry. Latterly he had been the president of the Scottish Mineral Oil Association and in that role he had taken the lead in matters affecting the industry generally. He led the employers' side in negotiations with the trade unions, and had been mainly responsible for arranging the contracts to supply fuel oil to the navy. He was a J.P for Linlithgowshire but took no great part in public life, being pre-occupied with business. He made a point of staying near Pumpherston for at least a few months in the summer every year. At his death he left a widow, four sons and four daughters.

The vacancy left by William Fraser's death was not easy to fill. His second son (also William Fraser), although only 27, had worked closely with his father since 1909 and had been a director of the Company since 1913. He was appointed joint managing director along with James Bryson, the general manager of the Company's Works.

During the Great War, the operations of the Company were much influenced by the government because oil was of vital importance to the war effort. In the years before the Great War the Navy was making increasing use of oil as fuel. For example, when Winston Churchill became First Lord of the Admiralty in 1911 he noted that 56 oil-fuelled destroyers had been built or were under construction. The shale oil industry was therefore of considerable importance to the navy and the Admiralty wanted to ensure the largest possible production of fuel oil.

Sulphate of ammonia was also important to the war effort. It was a good nitrogenous fertiliser and the Board of Agriculture made strenuous efforts to persuade British farmers to make full use of it. Its distribution eventually came under the control of the Ministry of Food. Another use of ammonia was in the manufacture of explosives, and so the government partly funded a plant at the Deans works to make concentrated ammonia liquor.

The importance of the shale oil industry in wartime was frequently acknowledged by government ministers and officials. In December 1916, Dr Cadman and Mr Houghton-Fry of the Ministry of Munitions met representatives of the shale oil companies and agreed that it was 'in the national interests that the production of shale oil should be as large as possible'. In January 1917, Dr Addison, the Minister of Munitions, emphasised the need for increased production. On 21

May 1917, William Fraser reported to the Board of the Pumpherston company that 'owing to the sinking of boats by enemy submarines, the Government find that it is desirable that the home production of fuel oil and other oils should be as large as possible and they are anxious that every effort should be made in the Shale Oil Industry to increase production.' Early in 1918 there was even a suggestion by the Admiralty that the industry should be taken over by the Government.

The Pumpherston company played an important role in these discussions with Government ministers and officials. On most occasions, the Company was represented by William Fraser, the son of the firm's founder, who also acted on behalf of the industry generally. These periodic meetings with the Government became more formalised as the war progressed. In January 1917 a committee was formed to oversee the industry. The five shale oil companies were represented on it, along with the Ministry of Munitions and the Admiralty. A separate committee represented the interests of the workmen. A neutral chairman was to preside over both committees and, if required, they would meet together.

Pumpherston Oil Works from the northwest, between the Wars.
The gasometer supplied gas for the furnaces.
(BP)

In June 1917, the Government decided to set up a department to deal with home oil production and Mr Fraser was asked to take charge of it. He declined but agreed to give what assistance he could, and this led to his being in London one or two days each week. Towards the end of the War, William Fraser was made a Commander of the Order of the British Empire in recognition of his contribution and that of the industry to the war effort.

There were significant increases in the cost of living due to the war and the men and their unions were anxious that wages kept pace with inflation. Various war bonuses were agreed with the companies and although there were occasional threats of industrial action, disputes were generally settled by arbitration tribunals set up by the Government conciliator, Sir George Askwith.

CHAPTER SEVEN

Scottish Oils Ltd

The Problems of Peace

DESPITE ALL THE PROBLEMS, the Great War was financially a successful time for the Pumpherston company and for the industry generally. In 1916, a dividend of 25% was paid; in each of the years 1917, 1918 and 1919, 40% was declared; and this despite the introduction in September 1915 of an Excess Profits Duty (equal to 50% of the increase in a firm's profits over the pre-war level).

The other companies reported similar results although on a less spectacular scale, and it seemed that the industry was in a strong position. However, in their reports to the Annual General Meetings in 1919, the directors of all four companies referred to the serious situation facing the industry. Proposing a dividend of 40% on the ordinary shares, the chairman of the Pumpherston company stated that although the results of the past year were satisfactory, most of the profit had been earned before the end of the War. He referred to the rising cost of materials and wages since the Armistice, and went on to say that 'should the Government find that they were unable to assist the companies and should the high costs of production and low prices for products continue to rule, it might be that the mining of shale and the manufacture of shale oil might become impossible, in which case it would be little short of a calamity to many connected with the industry. The Company had, however, a very efficient and valuable refinery which could refine very large quantities of imported crude petroleum, the refining of which is very much simpler than shale oil.' Similar statements were made by the directors of the other companies and the possibility of changing over to the refining of imported crude oil was to be brought up again and again in the severe difficulties faced by the industry during the 1920s.

This pessimism was partly due to the report of the Coal Industry Commission (the Sankey Report), which had recommended a seven hour day for the coal miners, and an increase in wages of two shillings per day. For years, shale miners' wages had been regulated by reference to those of the coal miners, and so it was assumed by the men that the Sankey Report would apply to them too. This would have cost the industry about £250,000 per annum for the wage increase alone. The increase in the price of coal which would be caused by the Sankey pay award to the coal miners would mean an additional cost to the shale oil industry of £215,000 per year, and this at a time when the industry was already operating at an annual loss of about £190,000.

Possibly more damaging in the long run was the effect of foreign competition. Throughout its existence the shale industry had been competing against petroleum produced from oil wells in the USA and elsewhere. This crude oil was of course much cheaper to produce than shale oil, and its production was expanding to cope with demand. Motor cars, buses and lorries were becoming increasingly common, and merchant shipping was converting to oil fuel. In 1914, world production of oil was 404 million barrels, of which the shale industry produced about two million barrels, about 0.5% of the total. By 1918, shale production had remained virtually static, while world production rose to 514 million barrels. In 1926, it had more than doubled. The shale oil industry was becoming an insignificant part of the rapidly expanding world oil industry, and the price of Scottish oil was determined by the price of imported oils. There was little scope to increase profits by increasing prices. In the years before the Great War, the industry had been able to survive, and at times prosper, partly because of profitable by-products, mainly sulphate of ammonia. In the post-war period the price of ammonia was affected by competition from a similar product from Sweden.

In May 1919, a meeting of the chairmen and managing directors of the oil companies agreed to send a deputation to London to seek Government assistance. The deputation stressed the ill effects of foreign competition, the increasing costs of production and the demands of the workforce for shorter hours and increased wages, all of which threatened the industry 'with a crisis which it can hardly surmount without assistance from the state'.

Formation of Scottish Oils Ltd

In addition to asking for Government help, it must also be said that the industry did not fail to seek other ways of reducing costs. In the years before the Great War this had been done through improvements in the retort, the basic equipment of the industry. Now the companies tried to cut costs by making changes in the structure of the industry. In the last years of the Great War the companies had been under some pressure from the Government to adopt common selling arrangements for their products. This had resulted in the formation of a new company, the Scottish Oil Agency Ltd, with a consequent reduction in the cost of sales. In July 1919 a similar joint arrangement was proposed for the manufacturing side. This was to be under the control of the Anglo-Persian Oil Company Ltd, one of the three largest British oil companies. The Government had acquired a controlling interest in this company in 1914, to secure Britain's oil supplies in the event of war. It was now proposed that by creating a new subsidiary called Scottish Oils Ltd, Anglo-Persian should take over the operations of the shale oil companies. The aim was to make savings such as had been achieved on the selling side by the Scottish Oil Agency. The Anglo-Persian company would also be able to supply imported crude oil to keep the refineries fully employed.

The offer by Anglo-Persian valued the Pumpherston company at £1,427,500; the Broxburn company at £470,000; the Oakbank company at £375,000; Young's company at £356,586; James Ross & Company at £168,750; a total of £2,797,836, of which the Pumpherston company made up slightly more than half. Payment was made in £1 Preference shares of Scottish Oils Ltd (i.e. shares on which dividend is paid before any is paid on ordinary stock). The ordinary shares of Scottish Oils Ltd, amounting to £1,000,000, were to be held entirely by Anglo-Persian.

The arrangement was not a straightforward take-over by Anglo-Persian. The five shale companies continued to exist, but their ordinary share capital had been acquired, apart from a small number of shares, by the new company Scottish Oils Ltd, in exchange for its 7% Preference shares. The Preference shares of the shale companies were not purchased and remained in private hands. The ordinary share capital of Scottish Oils Ltd was held entirely by Anglo-Persian. In effect, Anglo-Persian controlled Scottish Oils Ltd which in turn controlled the five shale oil companies. It should be remembered that the value of the pound was considerably greater in 1919 than it is today. The combined value of the shale companies, £2,797,836 corresponds to around £65 million at today's prices. Anglo-Persian had gained control of a very substantial business. In addition to the capital assets in mines and works, Anglo-Persian also acquired control of the substantial cash balances built up by the shale companies in the years up to and including the Great War, amounting to some £1,200,000.

By the end of September 1919, 98% of the Pumpherston ordinary shares had been transferred to Scottish Oils Ltd along with similar proportions of those of the other companies. The shareholders may well have been influenced by the Anglo-Persian's promise to underwrite the 7% dividend on the Scottish Oils preference shares until 1922. For the Pumpherston shareholders this was effectively a guaranteed dividend equivalent to 35% on their Pumpherston shares.

As far as the Pumpherston company is concerned, the creation of Scottish Oils Ltd means that the history of the Company and of the village itself is even more closely bound up with that of the industry as a whole. In 1919, the Pumpherston company was the largest and most successful of the firms in the shale oil industry. From 1919 onwards the Company ceased to be independent. It was controlled ultimately by one of the major companies in the worldwide oil industry and the operations at Pumpherston, Seafield, Deans and Tarbrax were affected by influences from far outside Pumpherston.

The directors of the new company, Scottish Oils Ltd, and of the five shale oil companies were nominated by Anglo-Persian. In fact the same five individuals were directors of all five companies, with William Fraser being the managing director in all cases. Fraser was the sole representative of the Scottish shale companies on these boards. James Bryson of the Pumpherston company became the general Works manager and Robert Crichton of James Ross & Co Ltd Philpstoun works, the general mining manager. Edwin Bailey, chief chemist to the Pumpherston Oil Company was appointed chief chemist of Scottish Oils in 1925. In 1923 William Fraser became one of the four managing directors of Anglo-Persian, responsible for production and UK distribution. In 1928, he was persuaded by Sir John Cadman, the chairman of Anglo-Persian, to move to London as deputy chairman of the Company. As a result of Fraser's move to London, James Bryson and Robert Crichton were appointed directors of Scottish Oils Ltd. James Bryson died in January 1930, having been associated with the Pumpherston Oil Company for 47 years. Fraser became chairman of Anglo-Persian (by then Anglo-Iranian) in 1941, and resigned as managing director of Scottish Oils and the various shale companies. Robert Crichton was appointed managing director in his place. Fraser was knighted in 1939 and created Baron Strathalmond of Pumpherston in 1955.

ROBERT CRICHTON

The fourth of the 'Big Five' oil company managers was perhaps even more committed to the welfare of the workforce than were his predecessors. Robert Crichton began in the engineering/mining trade in 1896 when he was thirteen years old. At the formation of Scottish Oils, he was appointed general mining manager, then managing director in 1941.

He was very active in public affairs, being among other things County Convener, an elder of St Michael's Church in Linlithgow, a member of the boards of St Michael's Hospital and of Bangour Hospital, a fellow of Heriot-Watt College, Governor of the Royal Technical College and a director of Grangemouth refinery.

He was an autocratic figure but respected by all, and with a gift for leadership. One typical and appreciated action occurred during the oil industry's most difficult years. Faced with the prospect of laying off a quarter of his workmen, Mr Crichton employed them all for three weeks in four and pulled strings to ensure they received a portion of unemployment benefit for the fourth week. There was not one strike during his 24-year reign and when he died in 1966 at the age of 83, his leadership of Scottish Oils had set an example to the rest of Scotland in the field of social welfare for workers.

Robert Crichton, managing director of Scottish Oils and County Convener of West Lothian County Council - 'a good man, a friendly kind of a man...'
(West Lothian Council Libraries)

However, the new arrangements did not solve the industry's problems, the principal one being the impact of the Sankey award on wages. The shale oil companies were convinced that the industry could not afford to pay a wage increase.

As early as April 1919, at a meeting of the Scottish Mineral Oil Association, the shale companies agreed to give notice 'terminating the wages agreement at the earliest opportune moment.' This naturally led to discontent among the workforce which came to a head in September 1919.

During the summer the Sankey award of two shillings per day was conceded by the employers, but the seven hour day for underground workers remained a sticking point. This had knock-on effects in the Oil Works, where the men wanted their working week reduced to 48 hours. The employers, believing that the industry could not afford it, refused, and the men voted to stop work on 27 September. The employers started to shut down the mines and Works in the preceding week; the retorts were cooled down and the refineries were stopped. However, in what was described as a 'twelfth hour settlement', the men agreed to accept the employers' offer that the present conditions and wages (including the Sankey award of two shillings per day) should be maintained until the end of the year. The plan was to find out whether the savings made by the amalgamation into Scottish Oils were great enough to warrant the concession of a seven hour day. The men also agreed to drop their opposition to the processing of imported crude oil. Eventually in April 1920, the employers conceded the seven hour day in the mines with a consequent 12.5% increase in piece rates.

However, this dispute led to a gradual closing down of the less profitable oil works, which continued until the middle of the 1930s. When work was restarted in November 1919, it was clear that not enough crude oil was being produced to keep all the refineries going. Scottish Oils Ltd therefore decided not to re-open the Young's company refineries at Uphall and Addiewell, unless crude oil could be imported.

CHAPTER EIGHT

The 1925 Strike

DESPITE THE DIFFICULTIES FACING the industry, the immediate post-war period had some successes. This was due, in part at least, to the amalgamation which was claimed to have resulted in a 20% increase in efficiency. In 1921 the five shale companies paid ordinary dividends totalling £120,000; in 1922, £63,000 and in 1923, £138,000. Most of these sums were of course paid to Scottish Oils Ltd, the holder of nearly all the ordinary shares of the companies. The position changed radically in 1924 when no ordinary dividends were declared, and only refunds of the wartime Excess Profits Duty enabled the shale companies to report small credit balances, after paying dividends to their Preference shareholders. It was clear that the shale industry was approaching a crisis. This is best illustrated by the fact that in 1914 the average end value of a ton of shale was 12s 6d, while production costs were 9s 11d – a substantial element of profit. In 1924 and 1925 the average value was 12s 5d, very little different from that of 1914, but the costs of production had risen to 14s 2d. The industry as a whole was operating at a considerable loss.

The workmen recognised the difficulties facing the industry at this time and made substantial contributions to its survival. In September 1923 they accepted a reduction in wages which substantially eroded the gains made in April 1920. The economies following the amalgamation meant that in 1924 only 7,500 men were employed, compared with 9,000 in 1919. In July 1924, came the first instance of work sharing. The Broxburn candle house was reduced to employing fifteen men, each of whom worked only one week per fortnight. Normally 21 men were employed in three shifts. The Addiewell candle house closed in 1923. Some of the mineral landlords also agreed to a reduction in rates of royalty (the money they got for every ton of shale mined from their land).

While the native Scottish shale oil industry was facing this difficult situation, Scottish Oils Ltd commenced the refining of crude oil supplied by the Anglo-Persian company. In 1919 the shale refineries at Addiewell and Uphall had been closed because of a surplus of refining capacity in the industry. Scottish Oils decided to adapt the plant at Uphall for the treatment of Persian crude oil, brought in from the port of Grangemouth at first by rail, but after 1924 by pipeline. At the same time a refinery was built at Grangemouth, also to deal with Persian crude. It began operations in 1924 and by August of that year was refining over 25,000 tons of crude oil per month – roughly equivalent to the total capacity of all of the shale oil refineries put together. In November 1924 extensions to the Grangemouth refinery were approved by Anglo-Persian.

In the autumn of 1924, the board of Scottish Oils Ltd was advised that if the situation did not improve it might be necessary to close down some of the less efficient works. In March 1925, things were made worse by the loss of the Admiralty contract for the supply of fuel oil to the Navy, despite appeals to government by both management and men. In September 1925 the management gave notice of the closure of Tarbrax works and mines; Dalmeny works and Ingliston mine; Broxburn crude oil works; the Dunnet and Newliston mines; and Broxburn refinery. It was also intimated that a reduction of 10% in wages at the remaining works and mines would be required to keep them in operation. The only alternative was to shut down the whole industry.

Complete closure of the shale industry would have had a devastating effect on the communities of the Almond valley. Over 7,000 men were employed directly and about 6,000 jobs in other industries such as coal mining, engineering, timber, etc, would have been affected by the closure. The threat therefore had the effect of bringing together representatives of various interests in West Lothian, Midlothian and Lanarkshire. A committee was formed which submitted a statement of the consequences of closure to the Prime Minister. The local MPs were also very active in arguing the case for the industry.

After a great deal of discussion and negotiation between management and the trade union as well as unsuccessful appeals for Government help, the crisis came to a head at the end of October 1925 when William Fraser, managing director of Scottish Oils, proposed to the union that, with a reduction of 10% in wages, the mines and works could be kept going as they were until March 1926, 'by which time I am sure we all hope that conditions will have greatly improved in our industry.' Mr Fraser stated that if this proposal was not accepted there was no alternative but a complete shut-down of the industry. In a subsequent ballot the men overwhelmingly rejected this proposal and went on strike on 10 November. The only hopeful sign in this gloomy situation was that the safety men at the mines and works were kept on.

A proposal for a Court of Enquiry was produced with the help of the Scottish Trades Union Congress but the men rejected this by a huge majority, because they would have had to accept the reduced rates while the enquiry took place. The main point at issue was the status of Scottish Oils Ltd. The men maintained that Scottish Oils and the shale companies formed a single entity, and so the profits arising from the refining of foreign crude oil at Grangemouth and Uphall could be used to offset losses in the shale plants, until conditions improved. The employers argued that the five shale companies remained in existence as separate concerns despite the fact that their ordinary share capital was largely owned by Scottish Oils. The employers' side prevailed. There was to be a 5% reduction in wages and the employers insisted that there could not be a full resumption of work

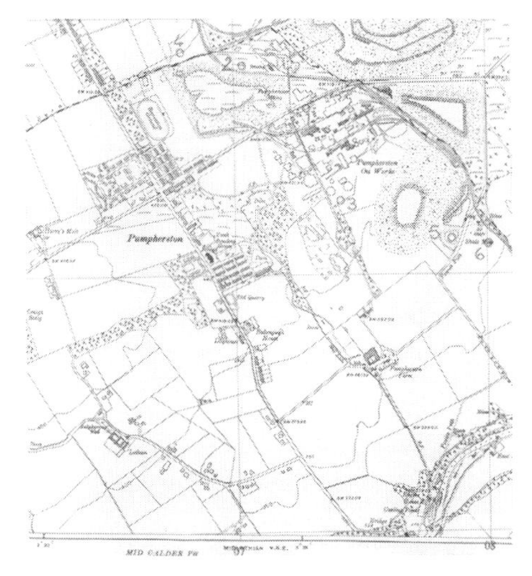

Pumpherston in the late 1940s. The circles show the locations of the six Pumpherston Oil Company shale mines. Numbers 1, 2 & 3 were abandoned in 1901, No. 6 in 1912, No. 5 in 1925 and the last, No. 4, closed in 1927. *(Ordnance Survey, Provisional Edition, 1952, six inches to one mile)*

at all the mines and works. This proposal was accepted by the men by a huge majority. The vote at Pumpherston was much closer although there was still a comfortable majority for acceptance.

Because of the time required to get the retorts back into working order, it was not possible to resume work until January and February 1926. Even then some works and mines were not restarted. At Broxburn the crude oil works and refinery, the Dunnet mine and Roman Camp Nos. 2, 3 and 5 Mines remained silent. At Dalmeny the crude oil works was closed. The Pumpherston company's Works and mines at Tarbrax and Woolfords Colliery were left standing idle. And in Pumpherston itself, the crude oil works did not re-start, nor did No. 4 Mine.

These closures meant a reduction of some 30% in the capacity of the industry. Until then, the shale industry had managed to maintain production levels very close to those of the prosperous years before the Great War. From 1919 to 1924, around two and three-quarter million tons of shale were mined and processed annually, only marginally less than the three million tons of the best pre-war year. By 1927, all hope of restarting the closed mines and works had been abandoned and the industry had declined to the extent that only two million tons of shale were used.

The strike and the delay in re-opening the mines and works naturally caused great distress throughout the shale area. The effect on the village of Pumpherston can be gauged by the fact that in November 1925 a number of employees of the West Calder Co-operative Society's branch had been laid off because of reduced sales. A communal kitchen was set up in the old bakery and by the end of November some 600 adults and children were being fed. The kitchen was well supported; local farmers supplied large quantities of potatoes, and a Leith baker sent in pies and cakes. The Pumpherston Pipe Band played to raise funds in Musselburgh, Newtongrange, Dalkeith, Peebles, Selkirk and Galashiels. Concerts were organised and collections were made at football matches. When Hibs played Aberdeen at Easter Road, £25 was raised. In January 1926, a cheque for £22 was received from 'a few Pumpherstonians in Detroit, America.'

The closing of the uneconomic sections of the industry did not return it to profitability. In May 1927, the Annual Report of the Pumpherston company showed an operating loss of some £20,000. The reports were as gloomy at James Ross & Company and the Broxburn Oil Company. Despite the men's acceptance in October 1927 of a further cut of 5% in wages, the board of Scottish Oils was informed in January 1928 that 'Notwithstanding that the costs of mining and retorting shale had been brought to the lowest and yields to the highest that they have touched in recent years, the losses in operations for the month of November... amounted to £10,000 without charging depreciation.'

CHAPTER NINE

The Hydrocarbon Oil Duties and the 1930s

THROUGHOUT THESE DIFFICULTIES, the Government had refused to assist the industry, despite its obvious importance to national security in the event of war. In April 1928, however, the Government's need for additional revenue led to the imposition of a duty of four pence per gallon on light oils. The good news for the shale industry was that home produced oils were to be exempt from this charge, giving the shale industry a four pence advantage in price over imported oils. This advantage, which eventually rose to one shilling and three pence, was initially confined to motor spirit but was later extended to diesel oil used in road vehicles (DERV).

Although this was a considerable help and inspired some confidence in the future of the industry, it did nothing to alter the basic situation of the shale industry. With a production of about 200,000 tons of crude oil per year, the shale companies operated within a world oil industry dealing with nearly 200 million tons per year. Throughout the 1930s the shale industry had to match the prices set by the large oil concerns and also to suffer the effects of the 'dumping' of oil from the Soviet Union at very low prices. Despite all this, the industry was able to respond positively to the opportunity presented by the new duty. In September 1928, it was decided to erect at Pumpherston, at a cost of £100,000, four cracking plants which '...had been designed with a view to dealing with products from shale in order that the fullest advantage may be taken of the duty on imported motor spirit.'

This minor expansion however was offset by further contraction of the industry. In July 1931, the works and mines at Philpstoun were closed. During the 1930s further closures took place at Seafield and Oakbank so that in 1938 only five crude oil works survived: Niddry Castle, Hopetoun (Winchburgh), Roman Camp, Deans and Addiewell. The total throughput of shale was about 1,500,000 tons per year, little more than half of that in 1924.

The contraction had of course a significant effect on employment. The total number in 1925 was 7,500, a significant decrease from the 9,000 at the end of the Great War. After the closures of 1925, the workforce of 5,347 fell gradually to 4,537 in 1931. In 1932 a sharp decrease to 3,384 was experienced, but numbers rose to 4,226 in 1933 and remained at about this figure until 1938. This increase was due to the introduction of a 'spread-over' scheme in 1932, whereby large sections of the workforce worked for three weeks out of four. In the fourth 'idle week' they were able to draw unemployment benefit. By this means, more men were kept in work and the hardships of the economic depression of the 1930s were lessened.

The Brick Works, June 1954. Moulded bricks being loaded by women onto a bogey, before being steam-hardened in an autoclave.
(BP Archives)

Some of the displaced men found jobs in the Grangemouth refinery and at Middleton Hall in Uphall, which had become the head office and central workshops for Scottish Oils. Some men transferred to the Anglo-Persian refinery at Llandarcy in South Wales, and some shale miners got work in the coal fields. Nevertheless, there was a considerable amount of unemployment. The National Union of Shale Miners and Oil Workers claimed that there were 922 unemployed shale miners and oil workers in the area at the end of 1936.

Pumpherston had been particularly hard hit by the closures. In 1925 the mines and crude oil works had been shut down and in fact they were never restarted. In 1926 only 318 men were employed. As oil refining was concentrated at the Pumpherston Works, the number employed there gradually rose again, reaching 483 in 1930. There was a sharp drop in 1932 to 407, but with the introduction of the

THE BRICK WORKS

'I was a brick press operator. You took the bricks off the machine press, put them onto bogies. Then they went into the claves' (pronounced claaves) for a bit – maybe eight to ten hours – getting steamed to harden them, then they were brought out, put onto lorries and away. There was 1,000 bricks on each bogey, and we filled 23 bogies a day, that was 23,000 bricks a day. When they mechanised the lifting, the machine could only do 21 bogies a day. The machine built up a bogey at a time, but the women could do it quicker.

The manager, he was a right stickler for getting the bricks out. Now the claves was like a pressure cooker, you cannae open it before the steam's out. The manager, he was too hasty and he was aye wanting the bricks out and onto the lorries. There was a big wheel you turned to open the door of the claves. He opened it, and the pressure pulled all the bricks out, and he emerged with his hair standing up. Everybody else had got out the way, they knew what was going to happen, but he was burnt and was taen away to hospital. He couldnae say he wasnae well warned.

You worked with a lot of lime and you came out white. We used to wear rubbers (sandshoes) and ankle socks, and we'd be spotless when we went out in the morning, and came home clarty. The lime got in your throat and your nose, and your nose'd be bleeding with it. Your hair and your eyelashes were white with it. There was no protective clothing – not at first. Well, you had rubber gloves for your hands, but they didnae last a week. The shale got inside your shoes and cut your feet, and coming home, you'd have blood squeezing out your shoes at every step.'

THE COKING STILLS

'The two worst jobs in the place were the paraffin sheds and the coking stills. The men went in while there was still an awfy heat. They went to break up the coke with big pinchbars, and throw it out the opening. They got a higher pay – that was the only reason they were on that job.

You had to wear clogs in the stills because of the heat and you'd to wear clogs in the sheds for the oil just rotted leather boots. They had wooden soles with steel toe caps and heels. All you heard was the clatter of the men coming from the baths to the sheds. The Works supplied them. You didnae do them up tight, you got used to them, just shuffling around in them.'

'spread-over' scheme, 500 men were in post in 1933 and numbers remained at about that level until 1938.

During the 1930s the Company started to make use of the spent shale bings which had become such a significant part of the local landscape. The spent shale was made into bricks. The process was very different from that of making traditional clay bricks. The spent shale was crushed, and mixed with hydrated lime and water. The mixture was then fed into brick moulding machines and the bricks

Coking stills at Pumpherston Refinery, c1920. After the light and heavy oils had been removed, the residue of the oil was distilled in pot or coking stills. The coke was then broken up by workers using pinchbars and working in temperatures so hot that it was not unknown for their wooden-soled clogs to smoulder.
(Almond Valley Heritage Centre)

were dried by steam in autoclaves. The bricks were a success. Large quantities were used by the shale mines for stoppings and other work underground. They were also used in building work around the works and the villages owned by the companies. Many of the offices and laboratories at Grangemouth are constructed of this very distinctive pink brick.

Pumpherston Works made extensive use of the new brick, one of the most prominent being the workmen's baths, provided in 1937. About twenty years previously baths had been installed for the use of men employed in cleaning the coking stills, a very hot and dirty job. During the 1939-45 war a canteen was built for the workers.

During the 1930s the refining of the crude oil produced by the shale companies was concentrated at Pumpherston and this involved considerable changes to the Works. The introduction of the duty on imported light oils made it necessary to increase the production of such oils, particularly motor spirit. To this end four cracking plants were built at Pumpherston at a cost of £100,000. The extension of the duty to gas oil used in diesel engines (DERV) resulted in the introduction of a diesel distillation unit to maximise production and make the most of the duty relief.

Westwood and expansion at Pumpherston

After the troubles of the early 1930s, the industry returned to profit in the period from 1936 to 1938 when it became obvious that war was coming. In 1938 it was decided to increase the capacity of the industry by the construction of a completely new crude oil works at Westwood near West Calder, along with improvements at some of the other crude works. This of course involved some increase in refining capacity at Pumpherston. Overall, the proposals were to cost £800,000, of which £135,000 was to be spent at Pumpherston on two new crude oil distillation plants. The Works now had the capacity to deal with 200,000 tons of crude oil per annum.

CHAPTER TEN

The Final Decline of Shale Oil

THE SECOND WORLD WAR meant that further expansion at Westwood was never carried out and the industry's capacity remained as it was in 1939. The story of the shale industry from 1945 onwards is one of gradual decline. By 1951 the Deans and Hopetoun crude works had been closed leaving only Westwood, Niddry Castle, Roman Camp and Addiewell. In 1954, John Caldwell succeeded Robert Crichton as managing director of Scottish Oils Ltd.

JOHN CALDWELL

John Caldwell, grandson of a former Pumpherston mining manager James Caldwell, was an officer in the Royal Flying Corps during the First World War and studied mining engineering at Heriot-Watt. He became assistant manager at Pumpherston in 1923, moved to become assistant general manager of the Central and Eastern Group of Scottish Oils Ltd, became general manager of the group and by 1952 was a director. In 1954 he took over from Robert Crichton as managing director. He retired in August 1962.

He was a quiet reserved man with a sincere interest in industrial and social welfare and a genuine concern for all the surrounding communities. He and Mrs Caldwell were always available if there were village organisations which wanted them or local occasions which would benefit by their presence. A lifelong member of the Golf Club, he came to several Golf Club functions, including one just before this death when he told the company he could see the corner of the course with the (then) sixth green at the pond from his study window in Ballengeich. 'And, oh, how I envy you lucky blighters,' he concluded.

He and his wife also attended a dinner given by the newly formed Youth Club (which had a couple of successful years in the 1950s but never quite got off the ground). He chatted to everyone afterwards but as he wandered round the hall, he carefully examined window ledges and fittings for dust. His frown indicated the hall-keeper would have a bad Monday morning in the manager's office.

John Caldwell and his wife are still remembered with affection by the older villagers. He was the third generation of the family to be identified with the shale oil industry and died of a heart attack at Ballengeich House in June 1970.

In 1944 there is the first mention of English crude oil at Pumpherston refinery. This was the result of exploration and drilling by the Anglo-Iranian Oil Company. A small amount of oil was produced by wells at Formby in Lancashire and at Cousland near Dalkeith, but the main production was from wells in Nottinghamshire, at Eakring and Egmanton. By 1948, over 4,000 tons per month were being refined at Pumpherston. In 1956 Pumpherston was dealing with

A presentation by foreman boiler-maker, George Allan to Jimmy Wilson.
Left to right: J. Philben, A. Armit, J. Wylie, E. Armstrong, J. Nethercote, unknown,
G. Allan (foreman), J. Dickson, D. Lightbody,
J. Lyons, J. Wilson, R. Armit, W. Quigley, J.Murphy, J. Howie, J. Millar.
(West Lothian Council Libraries)

70,000 tons of Nottinghamshire crude oils and 72,500 tons of shale crude oil per annum. This illustrates very well the declining fortunes of shale oil. In March of that same year a memorandum states that '...production of shale crude oil will fall off gradually and may decline to say 50,000 tons per annum in 3/4 years time.' By January 1957, the throughput at Pumpherston of English crude oils exceeded that of shale crude. The English crude oils had become the more important source for the Pumpherston refinery.

Detergents

The other significant development at Pumpherston in the immediate post-war period was the construction of the Detergent Plant. It produced a liquid detergent from a product derived from the wax extraction process. The detergent was sold in bulk to other manufacturers but it was also bottled at the plant under the name 'Iranopol', later changed to 'By-Prox'. Although women had been employed in the Works offices for some considerable time, and a few women were engaged at the brick presses, the Detergent Plant was the first location in the refinery to give employment to large numbers of women.

Women labelling and packing bottles in the Detergent Works, c.1955. Originally marketed as Iranopol, By-Prox was a liquid detergent for domestic use, derived from the wax extraction process.
(Almond Valley Heritage Centre)

Closure and after

In 1928, the shale oil industry had been assisted by the fact that it paid four pence less duty per gallon than imported light oils. By 1951 this had risen to nine pence less duty per gallon and had been extended to include gas oil used in diesel-engined road vehicles (DERV). In 1951 the benefit to the industry was £760,000 at a time when the annual wage bill was £1,500,000. The preference was later raised to one shilling and three pence per gallon, and had become the most significant element in the financial viability of the shale industry.

Britain's entry into the European Free Trade Association in 1962 made it necessary for this favourable treatment of indigenous industry to cease and the preference was withdrawn. It was a calamity for the shale oil industry. The end came swiftly: in its editorial of 6 April 1962, the *West Lothian Courier* headline was 'The Final Blow'; six weeks later the whole industry had closed down.

THE SOAP WORKS

'I went to the cooperage at the Soap Works; it was alongside the golf course. When the Soap Works started, all the detergents went into wooden barrels. There was the fear that metal barrels would contaminate the soap. There was seven or eight coopers, but they didnae make the barrels from scratch, they repaired and re-built old ones. They bought in ex-Navy rum barrels, and old barrels. You could smell the sherry when you were steaming them clean.

Later it was metal barrels, especially aluminium – sometimes 200 casks a month exported to Australia. There were that many different kinds of detergents, some really strong ones, really strong, you had to watch because they could burn your hand. They also made a paste for industrial cleaning – it was awfy strong.

It was all men that made the stuff, the only women was them that done the bottling – glass bottles, then later plastic. There was three shifts on the process side that made the stuff, but no shiftwork on the bottling side mostly.'

'At first it was glass bottles – Iranopol. The Iranopol was what they cried the soap work then. You'd take a handful of corks in each hand, putting them on the bottles. It takes a wee while to get used to it. There was only three belts. You done everything by hand – lifted the bottles onto the machine for labelling. Later it was plastic printed bottles. Sainsbury's soap was in red, white and blue bottles, striped; yon used to go for your eyes.

The women worked on the bottling and packing. There was six machines – belts – and six of you at a belt. One was the machine operator, she changed the type of detergent, and there was two put the caps on the bottles, one hammering the caps down and two packing. The one hammering, she had a wee wooden mallet and she'd to hold each bottle and tap the cap to make sure it was tight on. If you happened to knock one down, down they all went like dominoes. On the belt, you swapped around – an hour feeding, an hour capping, an hour hammering, an hour packing – right round the belt. Twice round, that was your eight hour shift.'

Over a thousand men throughout West Lothian lost their jobs, but a number of factors prevented the closure being a complete economic disaster. Some of the younger workers got jobs at the newly opened BMC truck and tractor plant at Bathgate. Some were transferred to the BP refinery and petrochemical works at Grangemouth, and some eventually got jobs as employment grew in the new town of Livingston, designated that same year, 1962. Pumpherston workers were better off than most, since the Refinery remained open for another two years, refining English crude oil. This and the success of the Detergents Works ensured that some 450 jobs continued at Pumpherston.

The Detergent Plant and the wax extraction process were carried on using oils brought up from Llandarcy in Wales, and Pumpherston detergents were used for a variety of products, some domestic, but mainly industrial. They became important in dealing with oil pollution at sea. A blow-moulding plant was set up for the production of plastic detergent bottles. The Works operated under the name BP Detergents, but in 1989 was taken over by the Robert McBride Group. The last name under which oil-related work was carried out at Pumpherston was Young's Detergents, a fitting reminder of the founder of the Scottish oil industry.

Late in 1992, 100 workers were laid off. The firm blamed ageing plant at Pumpherston which would have needed substantial investment to bring it up to efficiency. This, combined with surplus capacity at their newer Lancashire plants, boded ill for the future. The firm decided to concentrate on retail markets rather than the industrial markets which were the mainstay of the Pumpherston Works. In January 1993 the firm announced final closure, with the loss of the remaining 158 jobs.

Since then, BP has cleared and cleaned the site, and in consultation with the local community, has found a new use for the land. This process of rehabilitation is explained in Part Four.

CHAPTER ELEVEN

Working Conditions

The Mines

LIKE COAL MINERS, SHALE MINERS suffered from bad working conditions. After about 1877, the shale seams worked were from four feet to ten feet thick, and so shale miners were spared the cramped conditions of very thin seams. However the shale miner had to face three considerable problems: poor light, water, and bad ventilation. At the working face the only light was the cap lamp, the 'sweet oil lamp' of the miner and his drawer. There was always a great deal of dust and often powder smoke, and it was at times impossible to see the roof or the sides of a working place. Another problem was water, and there were several instances of flooding. In 1907 the miners at Philpstoun No. 5 Mine toiled in partially flooded workings, and Young's Newliston Mine closed in 1897 after being inundated.

The third problem was ventilation. The extensive use of explosives in shale mines made it difficult to keep the air clear. Early methods of ventilation were crude and indeed dangerous. The first major accident in a shale mine was directly related to the ventilation system: the shaft at Starlaw shale mine near Bathgate served as both the only entrance and exit, and as a chimney for a ventilating furnace underground. In 1870 the shaft's wooden lining caught fire and, trapped underground, seven men died.

The 1872 Coal Mines Act also applied to shale mines, and made it compulsory for all mines to have a minimum of two shafts to prevent disasters like Starlaw. Also mechanically driven fans gradually replaced the 'cubes' or furnaces of the early years. When the Pumpherston mines were sunk in 1883-84, the ventilation was entirely by fans, although 'cubes' remained in use by other companies.

Explosives

As far as fatal accidents were concerned, shale mines were no worse than coal mines. Shale, however, was more dangerous to work and resulted in more non-fatal accidents. Explosions of firedamp and accidents caused by explosives were the main causes. The frequency of firedamp explosions was probably due to the stoop and room method of working, which often left large areas of 'waste' where the shale had been taken out but the seam had not yet closed up. During the settling of the roof to the pavement, cracks opened and could allow gas into the waste and into the working places. Often men were tempted into areas of 'waste' in search of loose shale or to perform bodily functions – there were of course no toilets underground – and their naked lights could very easily ignite accumulations of gas. The firemen were responsible for ensuring that the working places in the mine were free of gas before the men started work. The men were paid by piece rates so any time spent waiting for ventilation to clear a working place was lost time and money. It was not unusual for a fireman to allow a miner to enter a place with the instruction to 'gie yon place a bit dicht' before starting work. This meant that the miner was to fan any gas away from the face with a piece of screen cloth or a jacket.

Many more explosives were used in shale mines than in coal mines, and the Inspector of Mines frequently commented on the number of accidents due to this. In the early years black powder was carried loose into the mine by the miner who had to buy the powder, in most cases from the Co-operative Society (supplied by the gunpowder works at Camilty near West Calder). There are frequent stories of powder bought in blue bags and stored under the bed, but no recorded incidents of injuries caused through home-stored gunpowder.

Many accidents were caused by the crude fuses used in the early days. They were made by the miners themselves from pieces of straw filled with gunpowder. In 1882 a miner was killed when filling a straw with powder at his working place; a spark from his lamp had ignited the loose powder. Incidents such as that became less common after the introduction of a manufactured fuse made of a cord with a core of gunpowder, and after the Coal Mines Regulation Act of 1887 prohibited the taking of loose powder into the mines.

Another common cause of accidents was the firing of a number of shots at the same time. Sometimes the miner was unable to get clear before the first shot exploded. The use of explosives required a hole to be bored several feet into the shale, then the powder in its bag had to be pushed right up to the end of the hole by a wooden stemmer, with a copper binding. The use of the wooden stemmer prevented sparks which might prematurely fire the shot, but if the hole was slightly out of true, the wooden stemmer might not be rigid enough to get the charge to the end of the hole, and it was not unknown for miners to resort to other tools rather than drill a fresh hole. In 1897 an eighteen year old Pumpherston miner was killed by an explosion thought to have been caused by sparks from a steel 'pinch' he was using instead of a stemmer. Another problem was the treatment of misfired shots. When a shot did not explode, the miner might go back too soon to find out why. Sometimes the misfire had simply been due to a slow fuse and men were killed by the resulting explosion.

Other hazards

Another cause of injury and death was the fall of shale or stones from the roof and sides of working places. This could be because it was difficult to see the roof of a working place in the dusty and smoky atmosphere. Injuries could also happen when clearing a place after shots had been fired; loose shale often had to be 'pinched' down from the roof or sides and this was a frequent cause of injury.

Jock Findlay (left) and Hugh MacKerracher, Pumpherston Works plumbers, 1950.
(Colin MacKerracher)

The shale seams often lay at a steep angle to the horizontal and this meant that miners might be in constant danger of loose shale or stones falling down the steep workings. There were also the dangers associated with the constant movement of full and empty hutches in the mine.

All in all, shale mining, although not as dirty as coal, was an uncomfortable, physically demanding and potentially dangerous occupation. Between 1895 and 1911, 122 men were killed in shale mines. Of these, 23 were employed by the Pumpherston Oil Company at its various mines at Pumpherston, Seafield, Deans and Tarbrax, but the Company had a slightly better record than other companies in the industry.

Over the years legislation was brought in to tackle the safety problems. The use of explosives was made subject to regulation: the number of shots to be fired at one time was restricted to two, and miners were required to leave a misfire for 30 minutes before going back to investigate. Safety lamps became compulsory in coal mines and, although the shale mines were exempt from this requirement for some years, the introduction of rechargeable battery-operated cap lamps made a significant contribution to safety. Despite this increased attention to safety, the most serious mine accident took place late in the history of the industry. In 1947 fifteen men lost their lives when an explosion of gas caused extensive damage to the workings at Burngrange Pit near West Calder.

The Works

The shale oil works of the Lothians used in excess of half a million tons of coal in a year, and there were continual complaints from landowners and farmers about the effects on their trees and crops of the smoke and gases from the works. The men working there also suffered from the polluted atmosphere. Many of the processes involved, such as retorting the shale and distilling the crude oil, needed high temperatures. Patrick Gallagher, the Irish labourer, writes of being nearly smothered by the gases and of leaving work as 'black as soot'. He describes removing hutches of spent shale from the retorts as so heavy physically and so hot that in each shift he drank half a gallon of water. Not all of the jobs would have been so hard and hot, but most were physically demanding.

There were also some dangers involved in the Works. Oil is an inflammable material, so fires were inevitable, and in the later years the Pumpherston company kept its own small fleet of three trailer fire pumps. There was also a considerable amount of gas involved in the process and this was perhaps even more liable to cause fires. The retort platforms were over twenty feet from the ground and in the early days were not adequately protected by railings, so falls were not uncommon. There was a constant movement of hutches and railway wagons, and crushing injuries were frequent. Injuries reported under the Workmen's Compensation Acts included broken bones, burns and scalds, lacerations, bruises and sprains. Many involved the amputation of fingers, toes, or even limbs.

PARAFFIN CANCER

Pumpherston Workmen's Baths opened in 1937. After work the workmen washed, then put on their own clothes which were kept in these lockers.
(BP)

'The paraffin sheds was the deadliest place in the Oil Works. That was where the oil and the wax were separated out. A lot of men took paraffin cancer. It was the old paraffin sheds that caused the trouble, where the men were handling the wax more. The men at the filters had shirts with no sleeves – they had to put lanolin all down their arms before handling the stuff. You've no idea how black and filthy it was in the sheds, oil dripping everywhere. You could hardly see the lights because they were covered with oil. Sometimes the filter sheet burst and oil was spewing out of it, you'd to put bags over it to keep the oil down. Everybody in the paraffin sheds got examined four times a year by Dr Scott at Broxburn. And you got soap and towels supplied. There was a good bath-house.'

Between 1874 and 1914, thirty-eight men were killed in the various oil works. In 1879 three men were killed at Uphall by an explosion of gas. In 1880 one man was killed while uncoupling wagons. In 1881 a retortman was killed by a fall from the retort platform at Broxburn. In 1885 a man was killed at Pumpherston when he fell between wagons being shunted. This incident was particularly tragic because the man had only that day started work and his name was not known to the management. In March 1890 a boiler explosion at Pumpherston Works killed three men and injured two others

THE 1890 EXPLOSION

'There was... a terrible upheaval of masonry and earth... Bricks and stones were sent in all directions for distances of several hundred yards...' Some dropped through the roofs of the nearest houses, smashing furniture, but fortunately injuring none of the inhabitants. It was fortunate that the accident occurred at 10 o'clock on a Saturday evening, for if it had been an hour earlier when the men were still at work, the death toll would have been much higher. As it was, two men were killed instantly – Edward Dempsey (50) and John Scott (35) – and one died an hour later – Andrew Taylor (39).

The *West Lothian Courier* reported the finding of the bodies in gruesome detail: one body had been 'thrown with considerable force high into the air and alighted upon the top of the first bench of retorts about 100 feet away... one of his legs was blown off and his body fearfully mutilated.' The other body was found 300 yards away from the explosion 'in a standing position, with his body leaning against a hedge. A shocking feature about this case was that the brains of the unfortunate man were found at an engine-house midway between the scene of the accident and where his body was found...'

The next day, Sunday, brought a crowd of sightseers to Pumpherston - 'so great were the numbers that a constable and six of the men of the Works had great difficulty in keeping the people outside the gates.'

In more recent years, greater attention was paid to safety and although many fairly minor accidents were recorded, the level of fatalities seems to have declined. In 1937, a fire and explosion in process stock tanks at Pumpherston caused the death of a cracking plant attendant (and former Hibernian footballer), William Dornan.

PART TWO

Pumpherston Village

by

SYBIL CAVANAGH

Introduction

IN THE EARLY YEARS of the shale oil industry, an American paying a visit to the district was asked what he thought of the village taking shape at Pumpherston. 'It will look grand', he announced after some consideration, 'when they get the hills finished.'

The 'hills' are finished now, if not quite in the way the transatlantic visitor imagined and although the district has never aspired to grandeur, it has certainly become presentable again. It is hard to believe that for nearly a century the bings, those immense accumulations of spent shale, were so much an accepted part of the landscape that they went unnoticed by the people who lived beneath them.

When the closure of the shale oil industry was announced, the *West Lothian Courier* paid tribute to the care which Scottish Oils had taken of its workforce. 'Scottish Oils Ltd pioneered in the field of workers' welfare, social and recreational facilities... Social and communal life reached a standard and strength unsurpassed possibly in the whole of Britain. It made for good citizenship and a high moral standard.' The oil companies did their best to create model villages for their workers, though it was not unrewarded altruism, of course, for the companies, providing both work and accommodation, were able to exert great control over the conduct of their workforce.

Probably the only people who can really appreciate the close ties that linked village and oil company are those who were brought up in the shadow of the old bings. The minor rules and regulations which governed their lives, understood rather than written, may give some idea of the relationship.

The tradesmen from the Oil Works, the joiners, plumbers and electricians, attended to repairs and maintenance in the village. This meant that minor damage such as blown fuses (the boxes were sealed) and broken windows had to be reported to the management and there were Works Office reprimands for the unfortunate miscreants, whether adult or child. All garden huts and outbuildings and the doors of all houses had to be painted in the same shade of dark green. Gardens had to be kept tidy. Misbehaviour by children out of school hours was followed by a note from the Works Office to the school headmaster. The letter named the guilty child, indicated the crime (most often raiding fruit trees or indulging in divot fights), and invariably ended with the rather ambiguous 'Kindly chastise this boy.'

Yet those who lived under the Company's regime betray no resentment of it and have happy memories of those days. Even allowing for the rose-tinting effect of nostalgia, it seems clear that the system was made tolerable because the men in charge were basically decent and well-intentioned. The managers were generally from a working-class background with practical experience of pits and oil works, and were therefore likely to have more sympathy with their workforce than most. Another point in their favour was that they did not abandon Pumpherston for some leafy suburb as soon as their means allowed. Even to the third and fourth generation, the management remained in Pumpherston, taking an active and knowledgeable interest in the village.

CHAPTER TWELVE

Pumpherston before the Shale Industry

Pumpherston Castle and the Douglases

PUMPHERSTON IS AN ANCIENT estate dating back at least to the 15th century; and the centre of the estate was Pumpherston Castle. A fortified tower house would have been needed both as a dwelling house and as a stronghold, for these were troubled times. The household would have withdrawn into the castle in times of danger, and their livestock would have been driven within enclosures for safety.

McCall's *History and Antiquities of the Parish of Mid Calder* states that the castle 'formerly stood in a field of about fifteen acres in extent, east of the present farm steading, which is still surrounded by a park [field] wall of stone and lime. The south-east corner of this field, though now cultivated by the plough alone, has always possessed a marked degree of fertility, and is regarded as the garden land of the old castle. A dove-cot formerly stood within the same enclosure, in front of the farm-house, and an ancient keep or look-out tower occupied a situation at the top of the bank rising from the River Almond.'

The site of the castle is believed to have been the field immediately north-east of the present farmhouse – probably about the middle of the field, though only an archaeological excavation could determine its exact location. Pumpherston Farm is well provided with dry stone dykes, some of them made partly of rough field boulders, and partly of cut stone which may well have come from the old castle of Pumpherston.

The castle probably fell into ruin when the estate ceased to have a resident landlord in the eighteenth century. The Rev. James Wilson, writing in the *Old Statistical Account* in 1791-3, states that in the parish of Mid Calder, 'There are no old castles in the neighbourhood which are capable of being inhabited'. The *Ordnance Survey Name Book* of the early 1850s concurs, stating that the castle 'was intirely removed upwards of 60 years ago' (i.e. before 1790). The Rev. John Sommers, however, in his 1838 account of the parish, says that 'the other very ancient building at Pumpherston... has long been in ruins and has lately been entirely removed'. The differing accounts can only be reconciled by assuming that some fragment of the castle survived into the early nineteenth century, and was then cleared away before 1838. Nothing now remains of Pumpherston Castle.

Various suggestions have been made as to the meaning of the name Pumpherston. One writer suggested it was from 'pamper', a short thickset man; another suggested it was from 'pundler', the official in the middle ages who

This aerial photograph taken in 1992 shows crop marks in a field east of Pumpherston Farm. (Crop marks appear above a ditch, or a wall or foundation. The greater or lesser moisture below causes a variation in crop growth, visible from above.) These marks reveal one rectangular and two circular enclosures which may have been some of the steadings and enclosures around the castle. *(Crown Copyright: Royal Commission on the Ancient and Historic Monuments of Scotland)*

impounded stray cattle. The most likely derivation is from 'ap Humphrey' meaning son of Humphrey. Pomphray was probably one of the Flemish (Belgian) noblemen invited by King David I and his grandson Malcolm IV to settle in Scotland in the twelfth century, and given grants of land in return for military service. This agreement was the basis of the feudal system, whereby all land was perceived as belonging to the Crown. The king could feu it to a baron in return for military service, and the barons could then sub-feu their lands to others, again in return for military service (later transmuted to money payments). The original owners retained some rights (including the right to feu duty) over the land even after it was sold.

Pomphray would have been granted the lands north of the Almond in return for serving the king in battle when required, administering the land and gathering royal taxes. Around the castle built by Pomphray, probably a wooden structure later replaced by a stone building, would have grown up a little settlement and farm to house and feed his adherents and servants – Pomphray's town – which over the centuries became Pumpherston.

Pumpherston was part of the barony of Calder, and so its feudal superiors in earliest times were the Earls of Fife, and later the Sandilands family of Calder House. The earliest known owner of the estate is William, Lord Graham, mentioned in a charter of the Great Seal of 1430. It may have been he who built the castle, or it may have been the Douglas family who by 1499 had acquired the lands of Pumpherston. The story of the Douglases of Pumpherston is well told in McCall's history of Mid Calder.

James Douglas, son of William, was the last of the Douglases of Pumpherston. After his death in 1696 or 1697, the land was sold to Alexander Hamilton, bailie (a kind of factor or law agent) of the Earl of Buchan's estate of Strathbrock or Uphall. He purchased both 'the manor place and policies of Pompherstoun and the fourth part of the lands of Pompherstoun and Knightsrig with the mains and mill thereof, and the lands called Forth, Bankhead, Backsyde and Muirhouse, and also the Kirklands of Livingston called Cannolands...' (i.e.

JOSEPH AND WILLIAM DOUGLAS

Two of the Douglases of Pumpherston deserve special mention, being colourful characters. Joseph Douglas inherited the estate of Pumpherston from his father about 1580. When a young man, he appears to have given offence to the daughter of a neighbouring landowner, Eupham McCalyean, only daughter of Thomas McCalyean of Cliftonhall. In 1590 she was executed for witchcraft, one of the charges against her being that some eighteen years ago, she had consulted Jonett Cwninghame in the Canongait, 'ane auld indyttit witch of the fynest stamp', about poisoning Joseph Douglas of Pumpherston by a potion of 'composit watter in ane chopin stoup' (i.e. a two-pint vessel).

Joseph's son William had taken up the offer of land in Ulster as one of the first batch of Scots colonists there in 1609 – one of the roots of the later Troubles. He was granted 2,000 acres, but in spite of this wealth – it was a far larger estate than Pumpherston – he did not settle there, but was back in Scotland by 1614. The Irish connection however was maintained for some time, as a document of 1658 records a contract to supply 60 cows to Douglas of Pumpherston's lands near Londonderry.

In 1634, William's wife, Isobel Sewart, made a complaint against another William Douglas, a relative who leased some Pumpherston land from her husband. 'On 29th June last, William Dowglas in Pomferstoun came by way of hamesucken [an attack in the victim's own home] to the complainer's dwelling house and at the instigation of Marion Bruce, his mother, entered the same about 10 o'clock at night when the complainer was alone, and, without any offence given by her, 'shamefullie patt hands in her person, she being great with chylde, gave her manie bauche and blae straiks in the face with his falded neiffes [fists], dang her to the ground, strake her with his feete on the bellie, rugged out the haire of her head, and thairafter drew his dager and sword and had not failyied to have slaine her thairwith if her husband had not happielie come for her lyfe; and then the said William flew.' The Douglases were a lawless family, for her own husband William Douglas had been up on a similar charge of hamesucken against Johnne Wricht at the Bridgend of Calder, some eighteen years before.

Isobel Sewart was as forceful a character as any of the Douglases. On 22nd September 1644 she was cited to appear before the Kirk Session of Mid Calder for 'scolding and railing against the Session, and was charged also with having said there was four hundreth merkis gott in fra the witches, and that the Session lieved [lived] thereon. This she denied, but said the minister had done many things behind folks' backis which he durst not do befoir their faces.'

Canonlands, later Carmondean). In 1704, he sold the lands of Knightsridge and 'Cannielands', but retained Pumpherston.

The Hamiltons held the lands of Pumpherston for two more generations, though neither they nor the Douglases were great noblemen. They were lairds or gentry, owners of a small estate – wealthy when compared with the mass of the people, and powerful locally, but not in national affairs. However the next two owners of the estate were great noblemen, purchasing the land and castle not as a residence, but as an investment.

Probably in the late 1750s or 1760s the estate of Pumpherston was acquired by the Earl of March, who took the trouble of surveying it together with his other local properties of Kilpunt and Illieston – presumably with the intention of making some improvements; but in fact the estate was sold again c.1770 to one of the great Scottish landowners, the Earl of Hopetoun.

Pumpherston Estate Farms in the Eighteenth Century

By the eighteenth century, the boundaries of the estate are fairly clear. The road from Uphall to Mid Calder was the western boundary. The Caw Burn formed both the northern boundary of the estate and of the county – Pumpherston being in Midlothian until the local government reorganisation of 1975, when it was transferred to West Lothian. The River Almond formed the southern boundary, and to the east the Bank Burn formed both the estate and county boundary.

The Pumpherston estate comprised 400 acres altogether and was therefore not very large as landed estates go. William Roy's map of the Pumpherston area in the mid eighteenth century shows that it comprised small areas of cultivated land amid large areas of rough grazing and moorland.

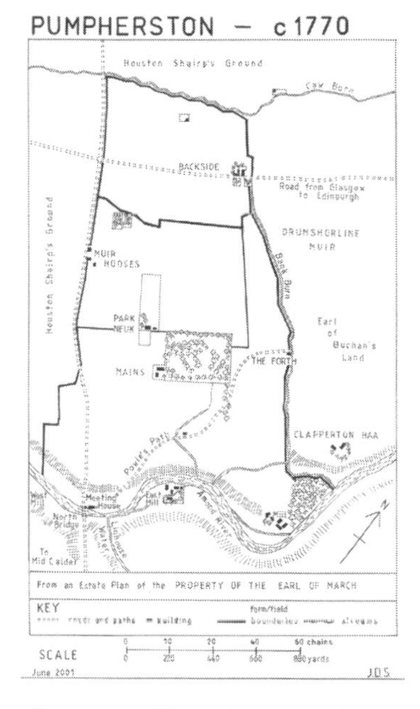

The estate of Pumpherston c.1770, before agricultural improvements swept away the farmtouns of the Forth, Parkneuk and Muirhouses. The name Forth derives from the ford where Powie's Path crossed the Bank Burn.
(Based on Register House Plan RHP 6739, by courtesy of the Marquess of Linlithgow, re-drawn by J.D. Slater)

POWIE'S PATH

'A footpath leading from Bridgend to Pumpherston, and for which some years ago a considerable cutting was made on the North bank of the Almond Water, during which operation, innumerable quantities of human bone were found, and a warlike weapon in the shape of a small sword, also stone coffins. The traditional accounts seem to indicate this River formed an important pass, and it was the scene of many conflicts between the Scots and Picts, similar remains being got at Bloom Park, Adam Brae, Cunningar (sic), indeed all along the banks of this River seems to have served as a cemetery for these unhappy victims.'

'The word 'Powie' signifies a skull. Some say it was the name of a general who had fell (sic) in action. This name may also have another signification which is – the word 'pow' being often applied to streams.'
Ordnance Survey Namebook, Parish of Mid Calder, 1850s.

The name is almost certainly derived from 'pow', a small stream or burn.

In the latter half of the eighteenth century, the farms on the Pumpherston estate were: Backside of Pumpherston – the old name for what later became Pumpherston Mains, whose site was built over by the brick works; Pumpherston (originally called Pumpherston Mains); and Muirhouses (later Harry's in the Muir), which at that time was *east* of the Mid Calder road. Each comprised some 90 acres.

Also part of the Pumpherston estate were a farm called the Forth consisting of 80 acres, and two smaller holdings – Pumpherston Park (also known as Parkneuk), which was an enclosed fourteen-acre field slightly northwest of the present Pumpherston Farm; and Pumpherston Mill which had attached to it 23 acres of land on the north bank of the Almond. (Until the eighteenth century, estate owners provided the grain mills at which their tenants and cottars were obliged to have their grain ground. This was known as thirlage, and from this comes the still common usage of being 'thirled' to something.) In addition to these farms, there was for a time in the late eighteenth century another holding called Little Pumpherston, which was slightly to the northeast of the old Pumpherston Farm cottages.

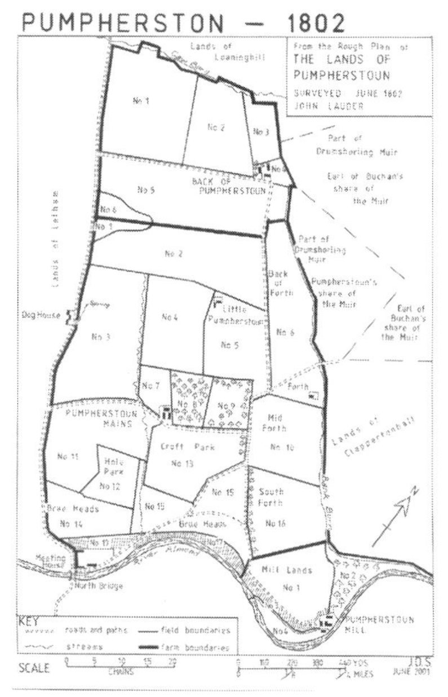

The estate of Pumpherston in 1802, after the Earl of Hopetoun re-organised the estate into two large farms, plus the Mill Lands. Some of the field names recall the old farms. The triangle formed by Powie's Path and Drumshoreland Road was not formally part of the Pumpherston estate, but was the part of Drumshoreland Muir which Pumpherston estate's tenants could use as rough grazing for their livestock – a sort of Pumpherston Common.

(Based on Register House Plan RHP 6742, by courtesy of the Marquess of Linlithgow, re-drawn by J.D. Slater)

Making Improvements to the Estate

Until about the mid eighteenth century, agriculture in Scotland was in a fairly backward state; much of the land was marshy, stony, bare of trees, and unenclosed. Crop yields were small, livestock was puny and the farms were often farmed not by a single tenant, but by several families working communally to eke out a minimal existence and living in accommodation that was often shared with the livestock. But the eighteenth century was a time of agricultural improvements, when agriculture began to be studied in a scientific way, and new methods of stock breeding and fertilisers were introduced. Stony land was cleared, boggy land was drained, land was enclosed into square fields by hedges and dykes, and trees were planted to provide shelter from the wind and to beautify the land. Small tenants and cottars were dispossessed and their holdings consolidated into larger farms. It was part of a sort of Lowland clearance which was taking place all over central and southern Scotland at this time.

This Agricultural Revolution can be seen in microcosm in Pumpherston during the era of the Earl of Hopetoun. The old open runrig farming was replaced by neat enclosed rectangular fields, belts of tree plantations, and the incorporation of the small farm holdings of Parkneuk, Forth, Bankhead, Muirhouses and Little Pumpherston into three larger, more efficient farms – Backside of Pumpherston (later Pumpherston Mains) and Pumpherston Mains (later Pumpherston Farm); and the Mill Lands. The name Muirhouses survived but was thereafter attached to a farm on the west side of the Mid Calder road.

What prompted these changes was an increase in population, which created a growing demand for grain and meat. A significant rise in grain prices from c.1780 provided the incentive to bring more land under arable cultivation.

BOUNDARY STONES

In 1787, the Earl of Hopetoun owned Pumpherston estate, and the neighbouring Clappertonhall was owned by the Earl of Buchan. Knowing the exact boundaries of an estate was necessary before beginning any improvements, and the Earl of Hopetoun was very much an 'improving' landlord. A plan of the boundaries of the two estates tells us that 'Their Lordships ordered Piled stones to be fixed in the Ground, upon or nearly to the old March (boundary)... and that a Straight Line from stone to stone respectively shall be the March for all times coming. Accordingly there is 14 stones in Number fixed in the Ground, each of these is of Hewen Stone, and upon each side of each of these Stones, faceing to the Earl of Hopetoun's lands is marked the letter H, and upon the East side faceing to the Earl of Buchan's lands by the letter B.'

A Period of Neglect

After the Napoleonic Wars, prices for grain fell and agriculture entered a period of slump. Some of the cultivated land on the Pumpherston estate was allowed to revert to moorland. The estate's owner at this time (from 1803 to 1817) was Henry Erskine, a well-known, witty and popular figure in his day. A younger brother of the eleventh Earl of Buchan, he received his early education at the school in Uphall (now one of the outhouses in the Strathbrock Church car park). He became a lawyer and the foremost advocate of his time, serving two short periods as Lord Advocate, the senior government post within Scotland. However, being a Whig during a long period of Tory rule, he made little political impact.

Whatever may have been his virtues as a lawyer, a wit and a leading light in the campaign for Burgh and Parliamentary Reform, Henry Erskine was sadly lacking in interest in his agricultural land, which seem to have suffered from some mismanagement as he concentrated on his legal and political career. He purchased the lands of Clapperton and spent the years of his retirement in landscaping and beautifying the banks of the River Almond around his house of Almondell. Upon his death in 1817, Pumpherston estate passed to his older brother David, Earl of Buchan, who, though a noted agriculturalist in his day, was by this time an old man intent upon many other concerns. Upon the Earl's death, childless, in 1829, Pumpherston passed to the twelfth Earl, Henry Erskine's son. Although he was a gentle, well-liked person, he had neither the taste for agriculture nor the energy to set things right. When he died, his wife erected to his memory the little shrine next to the Canon Hoban Hall at Broxburn Roman Catholic Church.

Peter McLagan

The estate of Pumpherston was acquired by Peter McLagan (the elder) in 1842. The land was in a state of some neglect, which the Peter McLagans, senior and junior, set about remedying.

Peter McLagan senior is known to have been in Demerara (British Guyana), South America, when his son Peter was born on New Year's Day 1823. The family's money derived from the sugar trade, and is therefore likely to have been based upon slavery.

Peter McLagan junior was a third son and cannot have expected to inherit the estate. He took a practical education in order to earn his own living, but his elder brothers predeceased their father, and he became the heir. He received his early education in Peebles and Tillicoultry and went on to Edinburgh University. He was nineteen when his father bought the estate and Pumpherston became his home. His interest in agriculture started early, and by 1852 he was respected enough in agricultural circles to be offered the post of editor of the *Journal of Agriculture*. Although aged 29, he was still financially dependent on his father, and when the latter proved hostile to the idea, he declined the post.

He began to contribute articles to several agricultural journals, based on tours to study farming in France and Italy and on his practical experience at Pumpherston. A letter from this period shows that he was very much a practical farmer: 'I have an engagement on Saturday at 10; but if I can get away after that I may take a canter along to Dechmont; if not I shall be at home all day, so that if you call, you will be sure to find me somewhere on my farm...'

He took an interest in all aspects of farming including the less glamorous such as the sanitation of farm cottages. Among his many published articles are 'On the feeding of cattle with cotton-seed and cotton-seed oilcake' (1855); and 'Silos and silage' (1886). In 1862, he tells a correspondent that he has 'just read the notes on 'Stuart's Earth Deodorizing Privy'.'

In 1860, Peter McLagan senior died and Peter junior now 37 came into possession of the estate. The work of improving the lands of Pumpherston had already begun. When the McLagans acquired the estate in 1842, writes Peter McLagan, 'The lands were about three-fifths cultivated and under cultivation, but in very bad condition, and greatly in want of improvements and two-fifths partially and wholly uncultivated... A part of the two-fifths had been cultivated during the wars of Napoleon' (i.e. 1792-1815) 'when the price of wheat was very high. But when prices fell after that, they were allowed to go out of cultivation and were soon overgrown with furze, broom, heath and rushes; and the other part was never under cultivation.' Part of the trouble was that it was poor wet land, formerly part of Drumshoreland Muir.

'The improvement of the three-fifths was commenced about 1847-8 by draining, liming, manuring and deep-ploughing, the effect of which was marked not only by the production of larger crops but by their earlier maturity, so that the harvest was about a fortnight earlier than it used to be... By 1862, the whole was planted and converted into arable land by drainage and other improvements.'

The Experimental Farm at Pumpherston

With his interest in practical farming and the science of agriculture, Peter McLagan was doubtless very willing to allow the Highland and Agricultural Society of Scotland to lease some of his land for research into farming methods. In 1877 or 1878, the Highland and Agricultural Society leased ten acres of land from him with the intention of setting up an 'experimental station'. The experimental farm was under the direction of Dr A.P. Aitken, chemist to the Highland Agricultural Society, but under the day-to-day running of Mr Tod, Peter McLagan's farm manager.

The land was divided into forty plots of one-quarter acre each for the testing of manures. Different quantities and different types of manure in a variety of permutations were tried over the course of eleven years, and the resulting crops were carefully monitored and recorded. The findings from these practical experiments were written up by Dr Aitken and published in the Highland and Agricultural Society Transactions, providing the first reliable information on the use of artificial manures. The experimental station operated until 1889 or 1890.

Parliamentary Career

Running parallel with Peter McLagan's agricultural career was his political career. In July 1865, having brought his lands into order, Peter McLagan entered Parliament as an independent. Later he stood as a Liberal because he was in sympathy with the major policies of the great Liberal leader, William Gladstone.

The 1868 election was a very bitter one in which the two candidates (both Liberals) did not hesitate to descend to personal insults and dirty tactics. Peter McLagan won and thereafter his seat was safe. He was an active and hard working MP. No less a person than the future Prime Minister Lord Rosebery said of him: '... there was (not) a more useful, or more practical member of the House of Commons than Mr McLagan. He is listened to when far more ambitious speakers would fail, because he has found out the golden secret of oratory, which is not to speak unless you have something to say.'

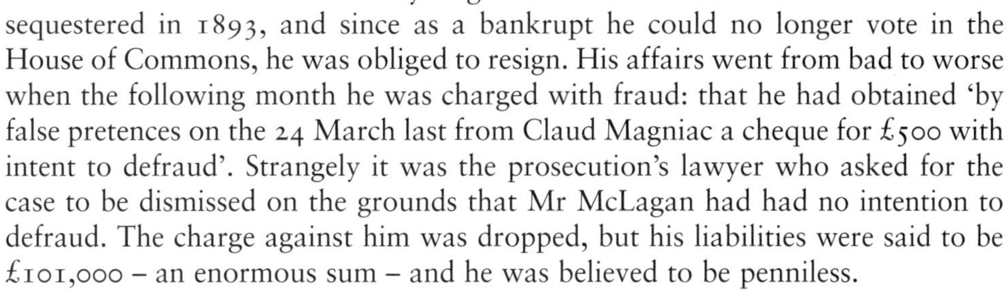

Peter McLagan, MP for Linlithgowshire 1865-93, and agriculturalist. *(West Lothian Courier)*

Many of the issues which he espoused in Parliament have become obscure to us now. He promoted legislation affecting land holding, agriculture, and the Game Laws, but he was particularly interested in temperance and 'even his opponents could not help admiring the dogged persistence with which he stuck to his pet scheme.'

His standing was high in the County of Linlithgow and in the House of Commons, but when he was over seventy he got into financial difficulties. His estate was sequestered in 1893, and since as a bankrupt he could no longer vote in the House of Commons, he was obliged to resign. His affairs went from bad to worse when the following month he was charged with fraud: that he had obtained 'by false pretences on the 24 March last from Claud Magniac a cheque for £500 with intent to defraud'. Strangely it was the prosecution's lawyer who asked for the case to be dismissed on the grounds that Mr McLagan had had no intention to defraud. The charge against him was dropped, but his liabilities were said to be £101,000 – an enormous sum – and he was believed to be penniless.

A meeting was held in the County Hall, Linlithgow, in July of that same year, when the great and good of the county decided to raise a fund in the county (£5,000 was the target sum) with which to buy an annuity (pension) for Mr McLagan. It is clear that Peter McLagan was well liked and respected, even by his political opponents, who were in fact the prime movers of the scheme. For the last few years of his life, he lived quietly in London. In 1900 at the age of 77, he died and was buried in Mid Calder churchyard.

Peter McLagan's period of 28 years as MP for Linlithgowshire has been exceeded only by that of the current Linlithgow MP, Tam Dalyell. Several McLagan sports trophies and the McLagan fountain at the Steelyard in Bathgate remain as a memorial of him, though the fountain was in fact donated to Bathgate by his wife.

Pumpherston estate was inherited by Peter McLagan's cousin, Charles Gibson of Pitlochry, who added the name McLagan to his own. The McLagans were landed gentry, not hugely wealthy, but comfortable. The estate has now been broken up, and although the McLagan Estates still own a little property in the Pumpherston area, the family is understood to live on the Isle of Wight.

BALLENGEICH

'On Drumshoreland Moor, within the grounds of Pumpherston Oil Company, there is a stone, popularly styled Bucksides – its correct designation being Backsides... This stone, a huge whinstone boulder about 12 feet long and 8 feet broad, was blasted in 1888, to make room for the site of a bench of retorts; a few fragments of the stone however yet remain by the roadside. The ancient name of this stone was Ballengeich – apparently the Gaelic for 'the township towards the wind', - as if a croft once stood here, near Pumpherston Mains, in an exposed and windy situation. Tradition, at any rate, avers, that round this stone in days gone by the Broxburn folks, along with their neighbours, used to assemble at Fair time, in the month of August, in order to witness their favourite sport of horse-racing; but whether there was any more ancient custom associated with it, we have never learned.'

Strathbrock, or, The history and Antiquities of the Parish of Uphall, by the Rev. James Primrose (1898).

Another writer claims the name Ballengeich derives from the 'gudeman of Ballengeich' – a name used by James V, who is said to have visited the spot when hunting on Drumshoreland Muir. There are, however, several places called Ballengeich, and the one associated with James V is probably that near Stirling.

Doghouses

The oldest building in Pumpherston is Doghouses, a fragment of which can still be glimpsed among the trees and undergrowth across the road from Ballengeich House.

Doghouses was built by Sir William Augustus Cunynghame, the owner of the Livingston estate from 1767 until his death in 1828. He was MP for Linlithgowshire from 1774 to 1790, active in public life and also sociable, hospitable and fond of hunting. In 1774, Sir William bought the farms of Letham and Craigs from the Shairp family of Houstoun. He was by this time the Master of the Linlithgowshire Hunt and built Doghouses on Letham Farm as kennels for the foxhounds about 1775.

The published history of the Hunt tells that '...a room is pointed out as being that in which the huntsman used to dine. Here it was, the story goes, that the kennelman or feeder returning home from Mid Calder one night the worse of liquor, and entering one of the lodging rooms of the pack, was set upon and totally devoured, nothing but his boots and one or two fragments of his clothes being found on the following morning. But the kennelman's life was not the only sacri-

Doghouses stood parallel to the Mid Calder road and comprised a two-storey dwelling house, flanked by two single storey wings as outbuildings. Presumably the hounds were kennelled in the wings, while the Hunt's kennelman had the centre block. c.1910.
(West Lothian Council Libraries)

fice on this occasion, for the Hunt lost a considerable number of hounds, all those which were concerned in this unfortunate affair having been immediately destroyed.'

The kennels remained in use for some twenty years, but by 1797 the foxhounds had been moved to new kennels in Linlithgow. Doghouses became housing for farm workers. In the mid 1850s, it was described as 'a small dwelling house in good repair, attached is a smithy and small garden, they are on the farm of Letham, and occupied by labourers. The alternative name of Dog House is given.'

KATIE McLEAN

The last occupier of Doghouses is still recalled by older Pumpherston residents – an eccentric old woman called Catherine McLean. 'Old Katie, a feeble, shrivelled old woman with boots buttoning up to the knees, lived in what was known as the Dog House... A wall and some bits and pieces remain but it was not really habitable even in Katie's day although there was a roof of a kind over one end. The surrounding area, which she called her garden, was a wilderness of nettles and gooseberry bushes run wild.' She was harmless but children would cross the road rather than meet the poor half-crazed soul with her pram selling tea around the doors. After her departure about 1948, the property was allowed to fall into ruin.

Letham Farm and Holdings

Letham Farm was originally part of the Houstoun estate, then was sold to Sir William Cunynghame, owner of the Livingston estate, and afterwards to Lord Torphichen of Mid Calder.

By 1920, the farm was disjoined from the Torphichen estate and was owned by John Martin. Agriculture was in a period of depression, so to supplement his income John Martin let three houses (including Doghouses) as well as a plantation of trees. In the 1920s and 1930s, he leased land for poultry farms. In the early 1930s more land was leased to Pumpherston Golf Club for a golf course.

LETHAM WELL

A notable feature of Letham Farm in the eighteenth and nineteenth centuries was Letham Well, a mineral spring strongly impregnated with sulphur, so giving off a 'foetid odour, resembling that of rotten eggs...' It was used by those suffering from 'Scrofulas, Gravel (kidney stones), and sometimes cutaneous (skin) diseases. Many years ago this well was handsomely built and enclosed by Dr Lamond who acted as Surgeon at Mid Calder. I may here remark that the water is constantly used by the people of Letham Farm, who have become so accustomed to it that they can scarcely find the noxious taste...'

In 1935/6 a large piece of Letham Farm was purchased by the Department of Agriculture for Scotland, in order to set up smallholdings. Several thousand were set up across Scotland in an effort to stop the drift from the land to the cities, and to encourage the unemployed to set up their own small rural businesses. Each holding consisted of some five to ten acres of land to be used for raising vegetables, fruit, pigs or poultry; and a house built to a standard L-shaped design like the small-holdings at East Calder, Mannerston, Dechmont, etc. There were 22 of these small holdings at Letham, let at a yearly rent of £28 33s. Some of the tenants remained in possession for many years or until the encroachment of Craigshill. A few of the houses are still in existence, but none is still used for its original purpose.

CHAPTER THIRTEEN

The Building of Pumpherston Village

Early Industrialisation

BEFORE THE SETTING UP of the shale oil industry, the only employment in the area was agriculture. In 1841 the estate of Pumpherston supported 33 people in seven households on the two farms of Pumpherston and Pumpherston Mains. After the purchase of Pumpherston estate by Peter McLagan senior in 1842 and their extensive work on the estate, the number increased. In 1851, Peter McLagan employed 30 farm labourers, six masons, a cook, a housemaid and a groom.

By 1871, the population of the Pumpherston farms and farm cottages had almost doubled to 109, and two thirds of the workers were no longer farm servants, but shale miners and oil works labourers or coopers. This very evident change was due to the sudden growth of the shale oil industry. Having found outcrops of shale on his land at Stankards near Uphall, Peter McLagan entered into partnership in 1866 with Edward Meldrum (former partner of James 'Paraffin' Young in the Bathgate Oil Works) and George Simpson, to form the Uphall Mineral Oil Company. Their oil works was developed at Uphall Station and by 1872 the firm employed 289 men. Fifty houses were built for their workforce at Stankards Row (better known as the Randy Raws) just north of Uphall Station.

Newcomers to Pumpherston, drawn there by the prospect of jobs at the Uphall Works, already made up half of the Pumpherston estate's population by 1871. Most were from Lowland Scotland – Lanarkshire, Fife, Stirlingshire and the Borders. Four were from England (the splendidly named Moses Finkle, and his three English-born children); but a substantial minority were from Ireland. They were not however permanent settlers in Pumpherston. All of them had moved on by the time of the 1881 Census. Another nineteen Irish-born incomers had taken their place, and it is clear that at this point, Irish workers came, worked for a few years, then moved on, or back to Ireland.

The first effect of the new industry felt by the Pumpherston estate people was this influx of new workers. The second effect was less pleasant – pollution. As early as 1872, the River Almond was becoming 'most offensively fouled by the waste waters from paraffin oil works, which have spoiled it for almost every purpose... When the water is low, and the nuisance therefore is greatest, it is no longer drinkable by cattle – it is no longer useful for any domestic purpose – it is no longer serviceable even for sheep washing – it is no longer the good trouting stream which it used to be... These Discharges not only cover the stream into which they pass with an iridescent film of offensively smelling oil, but the tainted water is capable of permeating through the soil into wells, spoiling their contents for domestic purposes.' The writer of these reports however, specifically excluded Peter McLagan's firm from blame: 'The Uphall Mineral Oil Company have used means at considerable trouble and expense, to utilize their waste water...'

Pumpherston Oil Company and the building of the Village

The formation of the Pumpherston Oil Company and the construction of the first houses took place in 1883-4.

It requires a feat of the imagination to think what might have been the effect on a mainly rural parish such as Mid Calder, of a huge influx of workers, the sprouting of a very large industrial complex in their midst, and the despoliation of great swathes of the countryside. Glimpses of the development as it appeared to the people of the time can be got from the minutes of the Mid Calder Parochial Board (Pumpherston was part of Mid Calder parish) which oversaw health, sanitation and poor relief.

'That Gross Iniquity'

The first mention of the developments in Pumpherston comes in the minute of the meeting of 6 May 1884, when Mr A.C. Thomson, manager of the Pumpherston Oil Company requested Mid Calder Parochial Board to supply the Works with water. It was agreed that after meeting the needs of the parish, there would be plenty left to supply to the Company, even though it required some 23,000 gallons a day. The Parochial Board took a realistic view of what they could hope to make from the Company: 'It is simply a matter of L.S.D. with the Company and if they had to pay more for Mid Calder water than £100 – £120 for their greatest requirement, they have better schemes in view.'

This arrangement persisted until 1890, when the Pumpherston Oil Company intimated its intention 'to terminate the Water Supply at Whitsunday.' This appears to have been a ploy by the Pumpherston Oil Company to force down the price. At any rate, Mid Calder parish continued to supply the Works for another few years.

The Works were well supplied with water, but there was concern in the parish over the lack of a proper water supply to the new houses in Pumpherston and their lack of sanitation. Parish authorities were obliged to appoint a Medical Officer, usually one of the local general practitioners, who made recommendations to ensure the health of the parish; and a Sanitary Inspector whose responsibilities were drainage and sewerage. In December 1884, 'The Inspector reported

Ballengeich House, built by the Pumpherston Oil Company for its general manager, and incorporating the Company's initials in stone above an upstairs window.
(Helen Scott)

the Parish as being in a satisfactory sanitary condition, except Pumpherston Rows, the drainage and other sanitary arrangements of which are not yet completed.' The houses were not ready for the incoming workers, but such was the pressure for accommodation that they were occupied before the water supply was connected.

Two months later in February 1885, the Inspector of Public Health reports that 'the want of water for domestic and dietary purposes is much felt by the inhabitants of Pumpherston Rows, now become a village with upwards of 500 of a population. He was instructed to communicate with the manager of the Pumpherston Oil Company and urge the necessity of this important want being supplied as soon as possible.' By May, William Fraser, the managing director of the Pumpherston Oil Company, assured the Parochial Board that he had taken steps to ensure a decent water supply and that 'the thorough drainage of the village of Pumpherston Rows would be very soon attended to.'

By February 1886 however, the Company had still not supplied the village with a water supply from Mid Calder Water Supply District, and the Inspector of Public Health, Dr Walter Watson of Howden House, had lost patience: 'That gross iniquity of a want of good water supply continues at Pumpherston. How long is it to be?'

The Inspector was instructed to intimate to the Pumpherston Oil Company that 'if a proper supply of good water be not provided within 14 days from this date, a complaint will be lodged with the Board of Supervision.' Since the matter then disappears from the Board's minutes, it is to be assumed that this threat finally moved the Company to take action. A water supply of some 24,000 gallons a day was brought to the village from Corston Spring. In 1895 the village was connected to the Bathgate district water supply from Forrestburn Reservoir. The Oil Works continued to use water pumped from the weir on the River Almond at Bridgend.

Sanitation

The other health problem at Pumpherston was the sanitation: there were no indoor WCs of course, and the outdoor privies left much to be desired: 'There was much offensive odour – apparently this is due to their position, as the prevailing winds reach them only indirectly.' These insanitary conditions may have contributed to an above average number of deaths at Pumpherston in mid 1886.

The Pumpherston Oil Company requested that a County Police Constable be stationed in the village. He also acted as the village Sanitary Officer but since he was paid by the Company, he was not in a position to force his paymasters to do his bidding.

By 1887, the parish Sanitary Inspector was much happier with conditions in Pumpherston: '...there is a plentiful supply of excellent water and cleanliness is being better attended to... their ashpits are regularly cleaned, and the drains kept as clear and clean as possible.' Thus nearly three years after the building of the village, it had at last reached a state of cleanliness acceptable in those days. The Company however required regular reminders from the Sanitary Inspector that 'the Manure heaps must be removed once a week in future without fail.'

Health

When Pumpherston village was built, there was already at least one GP in Mid Calder, who also acted as the parish Medical Officer. His services were available to all – for a fee – though the parochial board paid for his attendance upon those on the poor roll, and for the medicines and nursing care he prescribed for the poor. Whether his prescriptions were always to the patient's benefit is a moot point. In 1896, we find the doctor visiting a patient 'in 86 Pumpherston suffering from valvular disease of the heart', and prescribing 'whisky, eggs, etc.' Perhaps not surprisingly, the patient died. His father was summoned, but did not recognise the body as that of his son, so refused to pay for his treatment expenses or funeral. Fortunately his more compassionate workmates had a whip-round and raised £3 – enough to defray the funeral expenses.

The Oil Works was a magnet for tramps (including a surprising number of women) and vagrant workers. It was a large employer so there was the likelihood of casual work, and for tramps sleeping rough there were plenty of warm nooks and crannies among the machinery in which to pass the night. In 1902, the Pumpherston policeman, Constable Dunn, called at the Inspector of Poor's office in Mid Calder, to report that 'a tramp named James Thomson was lying in Works in a dirty condition and apparently unwell.' The Medical Officer, having examined him and suspecting smallpox, ordered his removal to Drumshoreland Fever Hospital (opened 1889). However, it was discovered not to be smallpox, and he was moved to the poorhouse.

MINUS BOOTS

In the winter of 1901, the 'Pumpherston Police at 6am brought a man to Office as he had been wandering about Pumpherston Village minus boots'. The Medical Officer could not certify that he was insane, but he was certainly 'in an excited state'. He was discovered to be Alexander Fairlie, 38, an engineman, who had mysteriously disappeared from Tarbrax two days before. His father was summoned from Harthill, paid for the Board's outlay on his son, and took him away.

Poor Relief

The main business of the Parochial Board, however, was poor relief. Given that by 1891, nearly half the population of Mid Calder parish was concentrated in Pumpherston, the proportion of Pumpherston folk requiring assistance from the Board was much lower than expected – as low as three out of 33 in 1898. This was because the Pumpherston workforce was not a 'natural' one with a normal proportion of the aged and infirm. Like any new town, it attracted those of working age, with a preponderance of the young and active. For a time it had almost no elderly people at all.

The first Pumpherston resident to be admitted onto the Roll of Poor, was 'Jane McGregor or Innes, widow of Alexander Forsyth Innes, shale miner, accidentally killed in the Pumpherston Shale Mine on 16 June last [1884], age 23 years – 2 young children, residing at Pumpherston Rows'. She was given 4s 6d a week.

In the early days of Pumpherston, many of the villagers were not eligible for relief from the local Parochial Board. Admission to the Poor Roll required a 'residential settlement' – five years of continuous residence. Many of the early claimants, though living in Pumpherston, had to claim from their previous place of residence or their parish of birth. Maria Galloway, aged 29 with four children – 'widow of Robert Galloway, oversman in Mine at Pumpherston, who died on 9 March last from Concussion of Brain caused by a piece of shale falling on his head while at work' – was not admitted because she had lived mostly in Uphall.

Poor relief was dependent on a means test. The Parochial Board (and later the Parish Council) had powers to investigate other sources of income and there are cases of people being refused relief, or having it withdrawn because they were receiving help from their family, or had savings in the Pumpherston Co-op, or in a friendly society. This is evident from the case of Mary Bryce Nailon, left a widow with six young children when her husband was killed in a Pumpherston mine. She was getting 10s a week from the Parochial Board when the Board discovered that she had received £50 compensation from Pumpherston Oil Company in a court case (of which she claimed legal fees took half). Her allowance was promptly stopped. She managed to prove, however, that she had to pay off £24 of debt at East Calder Co-operative Society, and so was re-admitted to the Poor Roll.

Another oddity of the poor relief system, to modern eyes, is the distinction made between the 'deserving' and the 'undeserving' poor. The morals of the applicant were taken into account when deciding the scale of relief. 'Elizabeth Welch or Copeland, aged 23, born in Pumpherston Mains, with a 2-day old illegitimate child, applied for relief and was offered the Poorhouse.' Grace Simpson, however, a deserted wife with two children, was treated very differently: 'on account of this woman being highly respectable and well-doing, it was agreed to continue meantime the 3/- a week.'

The Valuation Dispute

The work of the Parish Council was funded by the rates, so the presence of the Oil Works should have been a financial asset to the parish through the increase in its rateable value. In 1898, Pumpherston Oil Company appealed against the valuation which had been placed upon their Works and claimed exemption from the Poor Rates (April 1898). They withheld their payment – the largest single payment in the parish – while the dispute lasted, putting the Parish Council into such financial straits that they had take out a loan. It was concerted action on the part of some of the oil companies: Livingston and Bathgate Parish Councils found themselves in the same position. By March 1900, Pumpherston Oil Company's arrears of payment amounted to £788 17s 7d – a huge sum – and the Parish Council was obliged to impose an additional assessment on the other occupiers and tenants of the parish. The Company must have been extremely unpopular at this point. In 1901, the case was decided in the Oil Company's favour, and they were granted a 90% abatement of rates for their Oil Works, and 15% for their mines. All the parishes in Scotland which included an Oil Works then met in conference in Edinburgh and agreed to petition Parliament to alter the ruling, but apparently without success.

One side effect of the dispute was that nobody from Pumpherston would come forward to serve on the Parish Council as long as it was in dispute with their employer the Pumpherston Oil Company. When Parish Councils had superseded Parochial Boards in 1895, Pumpherston had been given three members – the first time the village had been given a formal place in local government. The three Parish Councillors were James Stoddart, Alex Skene, and James Quinn: two at least, Stoddart and Skene, were Pumpherston Oil Company employees.

However, at the 1899 Parish Council elections during the Valuation row, no candidates came forward to represent Pumpherston ward, nor did any come forward until the dispute was finally settled in 1901. One single large company controlling both employment and housing certainly caused a loss of independence among its employees. At the elections in February 1902, William Pratt, Charles Thomson (a Company miner) and Adam Porteous (a Company engineer) were elected for Pumpherston Ward.

CHAPTER FOURTEEN

Housing

Company Houses

THE WORKFORCE AND THEIR families required housing, and on a rural estate like Pumpherston, there was very little. The twelve houses at Pumpherston Mains (roughly where the brick works was later built) had been occupied by oil workers some ten or fifteen years before and they were used again. However, far more were needed, and the Pumpherston Oil Company chairman noted it as a priority at the AGM of 1884: 'in November [1883] we placed contracts for the erection of 48 double houses of a very good class, so that we might be in a position to command a superior class of workmen.' These houses were completed in April 1884.

Outside a South Village house, c.1930. The porch contained a sink and w.c.
(A. Maxwell)

By the time of the census in 1891, the village population was 1,382, and there were 210 Company houses in the old and new rows. By the standards of the time, the houses were well built, and many of them survive and compare well with modern construction. The Old Rows consisted of 21 one-roomed houses (single ends) built back to back in two blocks; 36 two-roomed houses built back to back in six blocks; 24 two-roomed 'through' houses (room and kitchen, with the room leading off the kitchen, the two rooms stretching the whole width of the block) in four blocks. The Work Rows had 32 two-roomed through houses in four blocks. The New Rows were made up of 88 two-roomed through houses in ten blocks of eight and two blocks of four. There were also six three-roomed houses built for the foremen. In the main street, there were two five-roomed cottages for the managers of the Works and mines and at Erskine Place, one house of four rooms and one of two.

The blocks in the Old Rows were arranged in three rows, and stood at the corner of the main road and Drumshoreland Road. The spaces between the rows were laid out as drying greens and shared wash-houses were provided. The pathways around the houses were paved with brick, but not the roadways. Most of the houses in the outside rows had gardens. The Work Rows in Drumshoreland Road were built in four blocks, two on each side of the road.

Facilities were fairly basic. At first the houses consisted simply of the one or two rooms, with the entrance straight from the street into the kitchen. The water supply was by nineteen street standpipes or pillar wells – one for every eleven houses. There were no sculleries or water closets, but instead outdoor privies were provided as well as ashpits for household refuse, and the Oil Company employed two scavengers to keep them clean. In the years leading up to the Great War, the Company built small porches over the doors of the houses in Pumpherston and its other villages. As well as providing some welcome screening from draughts, these porches contained a sink with cold running water and a separate water closet. By 1914, only a small proportion of the Company's 967 houses in its villages of Pumpherston, Livingston Station, Seafield and Tarbrax, were without sculleries and water closets.

Overcrowding

With some 210 houses accommodating 1,359 people, there was inevitably some overcrowding. Present day standards of housing are very different from those of the late nineteenth century; at present a house is regarded as overcrowded if there are one and a half or more persons per room. When Pumpherston was built the criterion was more than three persons per room. At 3.18 persons per room, Pumpherston was overcrowded even by the standards of the time. But the situation was not uncommon. In Midlothian 22.5% of the population lived more than three to a room, and an extraordinary 36% of the West Lothian population. Both Midlothian and West Lothian had a huge housing problem which was not tackled until the Council Housing Acts in the years following the First World War.

The largest Pumpherston household in 1891 was that of Irish-born James Kerr, who lived with his wife, six children and four Irish lodgers in two rooms at 5 Pumpherston Mains. Even more over-crowded was one single end which had ten occupants – husband and wife, six children and two boarders.

Boarders and lodgers were a common feature in the most overcrowded houses, and the Parochial Board of Mid Calder was aware of the problem. As early as 1887 it noted that 'several cases of overcrowding of lodgers have been found and partially checked'. Dr Watson the Medical Officer reported in 1888 that 'I have on several occasions visited the Miners' Houses at Pumpherston with the view to prevent overcrowding'. He was not very successful, judging by the 1891 figures. 106 households included lodgers – 44% of the total number of households; indeed 14% of the total population were accommodated as lodgers or boarders. (Boarders had bed and board; lodgers merely a bed.) Add to this the dozen or so households which included in-laws, uncles, brothers and nephews, and the number of households comprising more than just the nuclear family must have been almost one half. Today it is hard to imagine the sleeping arrangements or the lack of space and privacy in such a household, but there were strategies to cope. Box beds provided a measure of privacy, curtains across a room (or a blanket over a string) provided some separation of the sexes, and three or four children to a bed was the norm.

Yet despite these disadvantages, Pumpherston residents were fortunate in their housing – judged by the standards of the time. Compared with much of the housing in the shale areas, Pumpherston Oil Company housing was of a high quality, particularly after the improvements of the Edwardian period. In 1908, the Pumpherston Oil Company built five new houses beside the Institute, called appropriately enough Institute Place. A new row of houses was built about 1911 beside

PUMPHERSTON'S HOUSES IN 1914

'The Pumpherston Oil Company own 220 houses in this village, which is situated about one mile north from Mid Calder. These houses are built of brick, and rough-cast or cement-washed. In a number of cases, however, the back of the house, which faces the drying greens, is neither rough-cast nor cement-washed, and this presents an unfinished appearance.

There are about 173 houses consisting of room (12 ft x 11 ft, also bed recess and press), kitchen (14 ft 6 inches x 11 ft, also two bed recesses), scullery (7 ft 6 inches x 6 ft) with sink and w.c. A wash-house is provided for every four tenants. The rental is 2/6 to 2/9 per week, inclusive of rates. In the case of 36 of these houses, the room and kitchen are approximately 11 ft x 16 ft 6 inches each, and the scullery is 10 ft 5 inches x 7 ft 8 inches, which contains sink, set-in tub, boiler, and w.c.

There are 19 houses consisting of three apartments, and all conveniences as above, let at from 4/- to 4/9 per week. Then there are 27 single apartment houses, with scullery and w.c. let at 1/6 per week, inclusive of rates. Some 60 houses are built back to back.

A good supply of gravitation water is obtained, and a good drainage system exists. The refuse is collected daily by the Company, from dustbins provided to each tenant. The gardens are made good use of by the workers, and present a good appearance in the season.'

Housing conditions in the Scottish Shale Field: Evidence and Suggestions to the Royal Commission on Housing (Scotland), submitted by the Scottish Shale Miners' Association – March 1914.

the South Rows and was named Coronation Cottages to mark the Coronation of George v. The name was later changed to Pentland View, but older people continued to refer to them as The Cottages.

Private Houses

In 1899, the whole village could only boast of two owner occupiers: William Smail the blacksmith and William Paris, a mining contractor. Between them they owned seven houses, letting five of them to other tenants. Of the remaining 224 houses, 209 were owned by the Pumpherston Oil Company, and fifteen by the West Calder Co-operative Society.

Over the next fifteen years, just over a dozen private houses were built – Viewfield Cottage built for himself by Richard Cleland, an engine-keeper; and Marybank, owned by two Oil Works employees – James Ramsay, oil refiner, and Robert Petrie, chemist – and let as five separate houses. In 1906, James Caldwell, the Pumpherston Oil Company mining manager, though living in a Company house, built and let three houses. In 1906, David Fleming, a Pumpherston Oil Company foreman, built a block of six houses near the bowling green, and named it after himself and his wife Margaret – Davmar Place. An oil works labourer in 1891, Irish-born David Fleming had evidently prospered, and as was common at the time, put his money into house building, which yielded a regular income in the form of rents. His houses survive – the row of cottages to the south of the Store Corner, between the pensioners' houses and the two-storey house – but the name was lost when the terrace was renamed as part of Uphall Station Road c.1959.

Society Place

In 1891, the 63rd quarterly report of the West Calder Co-operative Society recommended the erection of a few workmen's houses at Pumpherston for their employees. By 1896, they had been built in two blocks and were occupied by Co-operative employees – a vanman, bakers, a shoemaker, butcher, and salesman. One of the fifteen houses was set aside for the village police constable.

One block was in Uphall Station Road, built gable end on a few yards south of the Co-op premises. The other was in Drumshoreland Road to the east of the Co-op. Some of the houses were two-apartment and some three. Sixty years later however, acceptable standards of accommodation had risen, and by 1960, at least half of the Co-op houses were uninhabited. They were later demolished and replaced by modern housing.

Housing between the Wars

By 1915, there were 234 Company houses, eighteen owned by West Calder Co-op and twenty privately owned or rented. During the First Word War, David Fleming sold Davmar Place to the Pumpherston Oil Company for £1,000 and invested the proceeds in War Bonds. In 1923, the Company built another twelve houses in Letham Park, bringing its total holdings in the village to 252. More must have been intended but were never built, for although the number of houses is twelve, they are numbered to 18. Privately owned and rented houses were reduced to a mere ten – in Marybank, Maryville, and Viewfield Cottage.

Pumpherston Gala Day procession passing some of the South Rows, c.1940s. The castellated parapets are typical of Pumpherston Oil Company design.
(A. Maxwell)

Eleven years later, in 1934, the situation had scarcely changed – 252 Scottish Oils houses, eighteen Co-op houses, and thirteen privately owned or rented houses. However, in the late 1930s, Midlothian County Council, busy with its council house-building programme to solve the twin problems of overcrowding and bad housing, bought land from Harrysmuir Farm and began to build council houses in Pumpherston.

GARDENS

'There were beautiful gardens then: I remember Harry Morris - he had a beautiful garden, his chrysanthemums used to win cups. He spent all his time in the garden because his wife wouldnae let him smoke his pipe in the house. If you didnae keep your garden neat, you were told. The Company told you.'

Council Housing

The first council houses to be built were those in Harrysmuir South and North, Terrace and Road, and the first to be occupied, just before the War, were in Harrysmuir South. Wartime restrictions prevented further building until victory was in sight. By 1944, the Council had acquired another site, and had 41 homes under construction – the pre-fabs at Harrysmuir Crescent (cheap and quick to build post-war). Then came Letham Gardens; and by 1955, Letham Grove and Letham Place. In 1955, with 57 families on the waiting list in Pumpherston, only twenty new houses were being built, so clearly the housing problem was not yet solved.

Even as late as 1954, Company houses outnumbered Midlothian County Council houses by 234 to 125. A mere seven houses were privately owned. Council house building resumed with Letham Road and Letham Terrace in 1957. The Scottish Special Housing Association had taken over some of the houses in Letham Grove by 1957/58 and in Letham Avenue in 1960/61.

HARRYSMUIR

'Harrysmuir – we used to call it Chinatown because the new street lights there made your skin look greeny-yellow.'

'It was a smaller village then – maybe not smaller in population, but much smaller in area. When I was at school, you could go round every house in the village and know who lived there, whereas now, there are houses not 100 yards away and I don't even know their names. But when I was a child it was the War years and that made a difference.'

By 1961, the improvements over the last seventy years were clear; the average occupancy in Pumpherston was just under one person per room, compared with nearly three persons in 1899. People who moved into the early council houses recall their mothers' pleasure in the space and the modern kitchens and bathrooms, though this was sometimes tempered by a sense of the loss of neighbourliness and familiarity.

Meanwhile Scottish Oils had not been idle. A process of renovation and upgrading of its houses was begun as early as 1940 and continued after the War. Its oldest cottages – the Old Rows – were demolished c.1961. An extensive modernisation programme was carried out with the assistance of Midlothian County Council. The Company gradually sold off its remaining houses to tenants, having generally knocked two houses into one to provide more spacious accommodation. The South Village retains something of its historic character, which, combined with modern comforts, now makes the houses sought-after properties.

Council house building continued in the 1960s. By 1964, Letham Avenue, Crescent and Gardens completed the Letham scheme; and Drumshoreland

Avenue, Crescent, Place and Road on the site of the old North Village or Old Rows had been added to the council housing stock, making a total of 306 council houses in 25 years. The Courier in 1963 commented favourably on Pumpherston's post-war re-development: 'the village must now rank in appearance, as one of the most modern in Scotland'. In the late 1960s, some 63 houses were built on the old Heather Wood site and were named Heaney Avenue, in honour of the long-serving Pumpherston councillor, Joe Heaney. The pre-fabs at Harrysmuir Crescent were substantially renovated in 1980 with pitched roofs, new external walls, kitchens, bathrooms and central heating.

The council housing programme had two important effects on the village: many residents for the first time enjoyed a house that met modern standards of size, space, sanitation and comfort; and council houses broke the monopoly of Scottish Oils, and gave locals a choice of employment. As council tenants, they were at last free to seek employment where they wished without facing the loss of their house. It also gave them what most of us take for granted nowadays - independence of their employer outside working hours.

Private House-building

During the first 50 or so years of the village's existence, Valuation Rolls show little advance in economic activity as measured by total valuation of the village. The remarkable fact is that while the oil companies ruled the village, Pumpherston remained the same size from c.1891 to 1931.

That there was so little private housing in Pumpherston seems odd, especially in view of the obvious need for more and the overcrowding of the early years. One reason may have been poverty. There were no large property developers or house-builders then, and new houses were mostly built by individuals as an investment. In Pumpherston there was perhaps a lack of spare funds among local people to undertake the task. Another reason was probably the dominance of the oil company. It already feued most of the land in and around the village from Pumpherston Estates, the landowners, and made no effort to encourage private building, over which it would have had no control. Almost the only development of private working-class housing was where some land was available – between Pumpherston and Uphall Station.

The proportion of owner occupation in West Lothian as a whole is 55.2% (1997). In Pumpherston, it rose from 21.5% in 1981 to 48% in 1991, the most recent figure available. These privately owned houses are mostly, however, the result of the sale of council houses to sitting tenants, rather than of private house-building. However, when the Bathgate to Edinburgh railway line re-opened in 1986, one of the effects was the boost it gave to private house-building in the neighbourhood of the stations. Edinburgh residents were attracted by the possibility of commuting by train to the city and by the cheaper property prices of West Lothian; private building boomed. In Pumpherston, Heatherwood Park was built in 1990-91.

A proposal to build perhaps as many as 900 new homes and a primary school in the Uphall Station and Pumpherston area over the next ten to fifteen years is under consideration by West Lothian Council. Local reaction is cautious – the benefits of improved facilities would have to be set against the increase in traffic and congestion and the loosening of community ties.

An unusual household was set up in Pumpherston in 1995, when 'Choices', a new 'care in the community' home opened in the village. Seven residents with special needs from the former Gogarburn Hospital, with the support of a care worker, share an ordinary home and a normal family life, and have been heartened by the support they have received from the people of Pumpherston.

CHAPTER FIFTEEN

Education

ONCE THE HOUSES WERE BUILT, the Company began to consider the other needs of its workforce – school, shops, and leisure facilities. Thanks to the Company, Pumpherston acquired more facilities than most villages of its size could boast.

A New School

For the first couple of years of the village's existence, Pumpherston children walked to Mid Calder for their schooling. The first move towards opening a school in Pumpherston came from the Pumpherston Oil Company, who in May 1885 wrote to the Mid Calder School Board offering to build a school for the younger children and to lease it to the Board. This was agreed early in 1886, and building went swiftly ahead. One room of this new school also served the village residents as a reading room. It continued in use until the opening of the Institute which had a purpose-built reading room among its facilities. A quiet place to read must have been hard to find in an overcrowded two-roomed house, so the reading room was a well-used retreat.

Opening day for the school was Tuesday 7 September 1886, and among the invited guests were various members of the School Board, and of course William Fraser of the Pumpherston Oil Company. Some 125 pupils enrolled and were all taught by the one teacher, Miss Marion Gray (annual salary – £65), with the assistance of a pupil teacher. Fortunately a second teacher was appointed within two months.

Class sizes were astonishingly high to modern eyes, though this of course was not peculiar to Pumpherston. With one teacher in charge of perhaps 60 or 70 pupils, discipline had to be extremely strict to prevent the class getting out of control. In 1888, with over 160 on the roll, Miss Gray had only two assistant teachers (i.e. c.53 per class). In 1891, with 284 pupils, there were five teachers (c.57 per class). In 1900, the HM Inspectors of Schools complained of 114 pupils at three different standards being taught in the one classroom!

The school roll expanded rapidly as the village grew – reaching 284 in 1891; and 370 in 1901. In 1903 it exceeded 400, reaching 455 in 1912; and did not again drop below 400 until 1917. After this period, however, both school roll and class size decreased, in line with a decline in the village population.

With so large a school roll, the School logbook is filled with complaints of overcrowding. The school was extended in 1888; and in 1890, temporary accom-

EMIGRATION

The fluctuations of the school roll were linked to the fortunes of the shale oil industry. When there were lay-offs, families would leave the village seeking work elsewhere. In prosperous times, there would be newcomers. However, the 1910s and 1920s were a period of marked decline in population, to which emigration certainly contributed.

In 1910, the *West Lothian Courier* reported that 'There seems to be a spate of emigration from Pumpherston at present, for quite a number of the male inhabitants are leaving for the colonies. Last Wednesday, the 'Peter McLagan' Tent of Rechabites entertained Bros. Alex. Simpson, Sam. Cameron and William Chapman, on the occasion of their departure for Canada. On the Thursday night a crowd of almost two hundred gave the emigrants a send-off at Uphall Station. Their intention is to seek work in the coal mines...'

modation was found in the Institute Hall. In 1902 it was extended once more, and yet again in 1911.

The problem of large classes was to some extent alleviated by the high level of absences; epidemic diseases were frequent, and absences for minor illness, spring cleaning, helping at home, and potato sowing or lifting, were commonplace, all of which must have brought some relief to the hard pressed teachers – while creating the problem of pupils at different stages of learning. In addition, school holidays were longer than today, and extra days were given for religious events, the twice yearly Mid Calder Fair and the various summer trips run by local organisations – Mid Calder Sunday School, Pumpherston Boys' Brigade, the Rechabites, Uphall Mission, the Band of Hope – some children managed to attend them all! Attendances markedly dropped on the day of Oatridge Races, Bathgate Procession, Orange Demonstrations, Broxburn Games and the District Agricultural Society Show. In October 1890, the school closed for the afternoon when the Prime Minister William Gladstone visited the Pumpherston Works.

The school was non-denominational and many Roman Catholic infants attended, then went on to Broxburn R.C. School when they were a little older. In 1896, East Calder R.C. school opened, and a number of Pumpherston children went there. Another few pupils were lost in 1897 with the opening of Uphall Station Infant School.

With its overcrowding and large class sizes, it is not surprising that in the first few years, Pumpherston Public School obtained only grudging praise from the School Inspectors: '...the general tone of the senior infant room is somewhat lifeless... mechanical tricks in slate work must be discontinued... Lavatory accommodation for the girls is needed...' In 1890 Miss Marion Gray became infant mistress, and a new head teacher was appointed – David Shaw.

He was a stern disciplinarian but an effective teacher, and under his leadership, the school at once began to obtain glowing reports, culminating in that of

Pumpherston Primary School, 2001
(Helen Scott)

1906: 'The present standard of attainment in all the classes of this school is so uniformly high that no report is necessary.' A wide range of subjects was taught, which even included Latin for the senior boys.

The school continued to receive consistently favourable reports – 'eminently satisfactory... thoroughly creditable... exceptionally well taught... very satisfactory indeed...' – up to and during the First World War. At this period, Pumpherston Public School was a very fine school indeed, and this despite constant changes of the junior staff. Most of the teachers were young women, and any who married had to give up teaching. Miss Marion Gray was replaced after twenty years as infant mistress by Isabella Nairn, and she in turn by Netta Horn in 1922. David Shaw however remained in post until his death in 1924, having completed 34 years at Pumpherston.

DAVID SHAW

'Mr Shaw was strict... he would belt you for looking at him. But parents backed the teachers then. If you got into trouble at school, your mother would just say, 'Well, you must have deserved it'. The standard of teaching was very high then. The head teacher and Miss Chisholm were both graduates, and at least two of the others...'

From 1903, James Bryson, the Works manager, presented gold watches to the boy and girl duxes of the school, and his widow continued the custom long after his death in 1930. For a time this was supplemented by silver medals presented by West Calder Co-operative Society.

The First World War brought new difficulties to the school. 'Miss Fergie absent on Monday as a Zeppelin raid result'. 'Mr Culbert, the woodwork teacher, was absent on active service'. 'Miss McGilp unable to attend the school this week owing to the death of her brother from wounds received in battle – the third and last to die in the War'. A half-holiday was given to mark the Armistice in November 1918.

Between the Wars

During the first 25 years of Pumpherston Public School's existence, schools were run locally by Parish School Boards, but funded and overseen by central government. Grants were raised or lowered according to the standards achieved. After 1918, School Boards were replaced by County Education Authorities and then from 1929 by County Councils – part of a general process of centralisation and greater state control of education.

At this period, Pumpherston was not a primary school in the modern sense. Those who were to continue to a secondary education went on to Broxburn Higher Grade School, Bathgate Academy or West Calder High School, and those who were not academic remained at Pumpherston till the school leaving age of fourteen. Pumpherston became a primary school on the opening of East Calder Junior Secondary School in 1940. Thereafter 'academic' children went to West Calder High School; 'non-academic' to East Calder.

Some bursaries were available for gifted pupils, for in 1920 two Pumpherston girls, Catherine Thomson and Annie Wynne, were at teacher training college; seven Pumpherston boys were undertaking courses in engineering and chemistry at Heriot-Watt College in Edinburgh; and John Shaw of 100 Pumpherston had a bursary to study medicine at Edinburgh University. These, however, were the fortunate few.

During the shale workers' strike of 1925-26 Pumpherston school children were provided with three meals a day, seven days a week, by the local soup kitchen: porridge and milk for breakfast; soup and bread for dinner; Oxo cubes and bread and butter for tea; and a meat pie for the weekend.

After his death in 1924, David Shaw was replaced by John Russell, the headmaster of Mid Calder School. In 1928 a two-storey schoolhouse was erected for him in Erskine Place, which had three bedrooms, a scullery and electric light.

The longest surviving members of staff during his tenure, were Miss Campbell, Miss Husband and Miss Chisholm who are remembered as being 'fairly

JOHN RUSSELL

Nicknamed 'Big Beefy', Mr Russell was a somewhat corpulent and unathletic figure although he had played senior football with Falkirk. He was rumoured to have suffered an unfortunate injury in his playing days which accounted for his rather stiff-legged gait. John Russell was a competent, if uninspiring teacher and an amiable man in a field that in those days tended to be populated by disciplinarians. He was the author of two highly regarded textbooks, *Standard English* and *An Approach to Standard English*. Under Mr Russell's guidance, annual school concerts were introduced, as well as Christmas parties, school sports and a gramophone record player!

agreeable provided they were not crossed'. There was a deliberate emphasis on handicrafts, particularly fretwork. The expectation was that only a few pupils would go on to higher education: the majority required an education that would fit the boys for a trade or for the mines and Works; and the girls for domestic service, then running a home. The boys were taught benchwork and technical drawing; the girls cookery, laundrywork, needlework and dressmaking. A school football team played occasional games against Oakbank and Livingston.

The longest serving teacher of all – Miss Elizabeth Anthony – retired in 1934

Pumpherston headmaster John Russell and Miss Chisholm, with a school class in 1945.
(J. O'Hagan)

after 41 years at Pumpherston. 1937 brought the excitement of the Coronation of George VI, to mark which the younger children were given a mug filled with chocolates and the older children a book on the Royal Family – and three days' holiday. In 1938, the first school outing was arranged – a trip by 68 pupils and seven teachers to the Empire Exhibition in Glasgow by special train from East Calder.

Vandalism, such an expensive problem for schools today, was almost unknown. Only three incidences are recorded in the Pumpherston logbooks down to 1960 – two break-ins in 1899 and 1929; and the theft of some lead from the school roof in 1953.

FAMILY LIFE

'There was a closeness then. Nowadays there's no family life. When the horn blew, the Scottish Oils horn, when it blew at twelve o'clock, everyone went home for their dinner, all the ones that stayed in the village; and when the horn blew at five o'clock, they all went home. There was no such thing as a latch-key kid. Your mother was always there, and she always knew what you were up to.'

The Second World War brought some changes to the school: during the black-out, school hours were altered to make the most of the daylight – a later start, an early finish and a shorter lunch-hour during the winter. Air raid shelters were built in the playground. One of the visiting teachers, Mr Andrew Steel, left to join the RAF. Brambles were gathered by the pupils and in 1941, 102 lbs of bramble jelly were made and distributed to local families. Over 60 lbs of rosehips were gathered and sent to an Edinburgh chemist to be made into rosehip syrup, a valuable source of Vitamin C. A few evacuee children attended the school, and vegetables were grown in a school garden.

Post-War

Mr Russell retired in 1949 and was succeeded by Mr William Mackay, whose first major problem was the inadequate heating of the school. The old open fires were replaced by stoves in 1952 without much noticeable improvement. The classrooms continued 'dreadfully cold' until 1955 when oil heaters were provided.

Mr William Mackay's departure in 1965, brought to a close a period of 75 years during which the school had only three headmasters. He was succeeded by Ronald McKay during whose tenure an extensive refurbishment of the school took place, a Parent Teacher Association was formed, and an exhibition on the history of the school was mounted.

Ronald McKay was followed by Mr McDonald in 1974. By then the prox-

WILLIAM MACKAY

William Mackay, previously a secondary school Maths teacher in Musselburgh, was an accomplished golfer, representing Edinburgh University, and a fervent rugby fan who never missed a Scottish International match at Murrayfield. He gave up the Pumpherston job in 1965 finding disciplinary matters irksome and feeling a need for the company of peers. He tried secondary school maths teaching again but did not settle to it and retired to Edinburgh where he died in 1971. A Highlander at heart, he was buried at Brora after a funeral service at his beloved Golspie.

imity and growth of Livingston New Town was affecting the village: the planners hoped to close Pumpherston school and send its pupils to one of the Livingston primaries. Parental resistance helped to safeguard Pumpherston School's continued existence, and in 1986 the school celebrated its centenary with a service, an open day and a commemorative mug for each child.

In 1988, Mrs Anne Kite was appointed, the first female head since Miss Gray a hundred years before. She was especially keen to involve the school in the community, and the community in the school. Her 1991-2 campaign 'Take Pride in Pumpherston' encouraged a responsible attitude to the local environment, and was commended at national and international level. The brightly coloured mural of Pumpherston's mining past on the wall of the Best Buy grocery shop is a lasting legacy of that campaign. Mrs Denise Armit succeeded her in 1993, and continues the good work.

In January 1975, Midlothian County Council opened a part-time branch library in a room in the Institute Hall under the supervision of Marjorie Lamond. Now, 26 years later, she is one of the longest serving employees on West Lothian libraries' staff.

The School and the Company

During the early years when the school was run by a School Board, considerable influence could be exercised by the Pumpherston Oil Company through its senior managers, Caldwell and Bryson, who were members of the Board. From 1919, however, the school was financially and administratively under the control of the local authority.

Few of the teachers actually lived in Pumpherston; most, including David Shaw, seem to have commuted by train to Uphall Station, and this may have helped them to avoid undue influence by the Company. Indeed, the school and the Co-operative Society were the only organisations in Pumpherston more or less independent of the Company. However, the non-residence of teachers, doctors and ministers in the village meant that there was a dearth of professional people, giving the Oil Company 'gaffers' a disproportionate influence on the life of the village.

CHAPTER SIXTEEN

Church

IT HAS OFTEN BEEN asserted that the Pumpherston Oil Company refused to allow either a pub or a church to be built in the village. The former has some basis in fact, as both Peter McLagan the landowner, and the Pumpherston Oil Company directors frowned on alcohol. That the Company refused to allow a church in the village is a myth. On the contrary, the directors encouraged church activities, as tending to produce a moral, hard-working and conscientious workforce. It was the Churches who failed to produce funds for the building of a church, which the Oil Company would have welcomed.

RESPECTABILITY

'When I was a child, my father could remember a young woman who threw herself off the viaduct – the railway viaduct down there – when she fell pregnant – not married, you see. As a child, there was no divorce, I didn't know a divorcee. They wanted to be seen to be respectable.'

When Pumpherston village was built, there were already two churches in Mid Calder parish which were near enough to be accessible to most Pumpherston residents. The closest church was that at Bridgend, on the north bank of the Almond.

The church at Bridgend traced its origin to 1761 when a congregation of the Associate Presbytery (a Presbyterian denomination which had seceded from the main Church of Scotland) began to meet in Mid Calder. The 'meeting house' was built in 1765 on the 'Bridge-haugh', part of the estate of Pumpherston. In 1847, various secession churches came together nationally to form the United Presbyterian Church, and the church at Bridgend was henceforth known as Mid Calder U.P. Church.

Even before the establishment of the Pumpherston Oil Company and its village, William Fraser while a manager with the Uphall Mineral Oil Company, was a member of the church at Bridgend and by 1873 was an office-bearer. Other Oil Company managers such as William Gray and James Caldwell were prominent office-bearers. In 1886, the Kirk Session recorded its indebtedness to William Fraser 'for the sympathy and interest he has manifested in the efforts they have inaugurated and carried out for the religious and moral benefit of the employees at the Works.'

The original Bridgend church was built on part of the Pumpherston estate, and stood gable end to the road. This second building (extreme right) was built in the mid nineteenth century and was demolished about 1974.
(Almond Valley Heritage Centre)

After another national church union in 1900, Bridgend became the United Free Church, and continued so till 1929, when the United Free Church and the Church of Scotland re-united 86 years after the Disruption of 1843. The parish of Mid Calder was divided into two, and the north part (which included Pumpherston) was in Bridgend's parish.

The Church was an important centre and source of social life, with its Woman's Guild, Sunday Schools, Young Men's club, youth groups, soirees and other church meetings. One of the best remembered ministers was the Rev. J.M. Jeffrey who came to Bridgend in 1925 and died there while still its minister at the age of 84 in 1951. The last Bridgend minister was the Rev. Fred Robertson, then St John's and Bridgend parish churches united in November 1956 to form the Kirk of Calder. The Bridgend church building was demolished in the mid 1970s.

For a short time, however, Pumpherston had its own minister. In 1905, Mid Calder U.F. Church applied to the United Free Church Home Mission Committee for a grant to place a minister in the village of Pumpherston. 'There are 230 families in the district, or a population of 1,600, composed of U.F.s, Established Church, and R.C.s. Meetings are held in a hall seated for 250 to 300. There is no other agency at work besides that now proposed. The working expenses will be met by the givings of the people, and the Pumpherston Oil Company are expected to contribute liberally to the missionary's salary.'

The Church's expectations were not disappointed. Pumpherston Oil Company, through William Fraser, its managing director, gave 'generous assistance' towards this scheme and also for the establishment of a mission station in Livingston Station and Seafield, promising 'all assistance' and assuring the Committee of his 'earnest desire to support the Church in its work'. (A Mission did not have full congregational status and was supported financially by the national Home Mission Committee. It was assigned a minister or a lay missionary, and if it flourished and became self-supporting it could apply for full congregational status. Some like Blackburn did achieve full status; Pumpherston did not.)

The Rev. James B. Macdonald was appointed to Pumpherston Mission at the beginning of 1906. A highly educated man of academic interests, he stayed for barely two years in Pumpherston, before moving to the full charge of Langholm in 1907.

He was followed by the Rev. J.F. McHardy, who probably related better to the mainly working class congregation than his more intellectual predecessor. Under Mr McHardy, various church-based groups flourished: a Bible Class, the Pumpherston Mutual Improvement Society (which included female members) and the Pleasant Sunday Afternoon Brotherhood (which didn't).

In 1910, Mr McHardy moved away and 'the mission work at Pumpherston being in abeyance, Mr [Thomas] Peden [of Broxburn] was asked to go out there, and see if anything could be made of it. He was successful in a short time in establishing the mission work there with the most encouraging results, especially among the young.' Under the supervision of Bathgate U.F. Presbytery, Mr Peden, a lay missionary, continued his work in Pumpherston until 1921.

At this point, the Mission Station of Pumpherston ceased. There were not enough U.F. Church adherents to make the Mission self-supporting – i.e. to pay the minister's salary and perhaps build a church – so Pumpherston's U.F. members reverted to the church at Bridgend.

In addition to the Mission Station work, a Sabbath School was established in September 1886 by Mr William Smail, an elder in the Bridgend church, who took on the work 'on condition that he does not organise the picnic'. Two months later there were 145 on the roll and sixteen teachers. Mr Smail died in 1909 and is buried in Mid Calder Cemetery. A later Sunday School superintendent, Davie Blain, foreman joiner at the Works, is remembered with affection in the village.

Since 1956 Pumpherston has again found itself within the parish of the Kirk of Calder. Many Pumpherston residents attend the parish church in Mid Calder but there is also a service every second Wednesday in the Pumpherston Institute Hall and a Sunday service there once a month, which is conducted by the present parish minister, the Rev. John Povey. Pumpherston, however, is the largest village in West Lothian not to have its own church building.

There has also been a long association between Pumpherston and the Roman

Catholic Churches in Broxburn and East Calder. Broxburn Roman Catholic Church served the R.C. communities of Pumpherston and the Calders until the building of a combined school and church (St Cuthbert's) in East Calder in 1896. Since then the Catholic community in Pumpherston has gone to East Calder for worship and education.

As for the pubs – they had to wait until the influence of the Oil Company waned after the Second World War. Before the 1950s, drouthy villagers had to walk to Mid Calder for a drink. In the mid 1950s, David McCreight bought a piece of land from Midlothian County Council and built the Sevenoaks Roadhouse. The Cawburn Inn was set up by John Ramage on the northern boundary of the village at about the same period. Both pubs are still flourishing and are important social centres for the village.

CHAPTER SEVENTEEN

Self-Help, the Co-op and Other Shops

BEING A COMPANY VILLAGE, Pumpherston had many facilities provided for it by the paternalistic care of the Company, which other communities had to provide for themselves – housing, jobs, sports and leisure facilities, and health care. There was less need for self-help than in a traditional community, but nevertheless there is some evidence of it.

In the days before the National Health Service, the threat of illness or injury weighed very heavily on working people. Medical care had to be paid for and if the illness prevented the wage-earner working, the family might find itself in grave distress. There were ways in which a community could insure itself against these difficulties: the most common were the formation of friendly societies and nursing associations.

Friendly Societies

Until 1897, workers laid off by illness, accident or infirmity had few means of support other than going 'on the parish' – i.e. applying to the local parish authorities for assistance. This was considered – and was intended to be considered – a humiliation, and most families dreaded such a necessity. But low wages meant that few men could save enough to support their families during a prolonged absence from work.

From 1897, the Workmen's Compensation Act gave the right to compensation for industrial injuries, regardless of negligence. This was a great step forward, especially in an industry in which one in ten shale miners suffered an injury in any one year. But compensation was often inadequate, and many workmen sought other means of insuring against the inability to work.

One solution was to join a friendly society. These were associations of working men, sometimes from one works or trade, sometimes of general membership, whose aim was to provide sickness benefits and funeral expenses for their members. Members put away an agreed small sum each week, and in return, the friendly society paid out an allowance to ill or injured members, or to their widows in the case of a death.

In Pumpherston there appear to have been several such societies. A Pumpherston Friendly Society was formed in 1895, but failed to flourish and no more is heard of it. It was a purely local society with small funds and could have offered no more than basic sickness and funeral benefits. By 1909, there was a Mining Society – presumably a friendly society – but no more is known of it.

National friendly societies such as the Shepherds and the Gardeners flourished at this time. They had much larger funds, the financial security of forming part of a national organisation, and the trappings of a secret society to attract members. Several of these 'Affiliated Orders' had branches in the area. The Rechabites (who demanded total abstinence from their members) had a branch (or tent) in Pumpherston by 1909, with a juvenile section superintended by D. Ronald. The Loyal Order of Ancient Shepherds had a branch in Mid Calder from 1887. By 1910, it had 769 members, many of whom were Pumpherston men. The Foresters had a small branch in Mid Calder, as had the Gardeners and the Rechabites, and among them they mustered 1,396 members in Mid Calder parish. It is likely that a large proportion of Pumpherston workers were members of one or more of these friendly societies.

In 1911, the National Health Insurance Act made insurance contributions compulsory. The monies were collected through approved societies, which included most of the national societies. They flourished and their memberships increased. Despite the success of the Affiliated Orders, there may still have been a need for a local society which asked for smaller contributions – perhaps as low as a penny a week – and had the further advantage of being 'yearly' societies, dividing up among the members whatever money was left at the end of each year.

Two purely local friendly societies were set up in Pumpherston. Some of the Oil Works employees set up a Benefit Society in 1918 with a weekly contribution of 4d. By the end of its first year, 293 workers had joined. In the following year, the mines department of the Pumpherston Oil Company decided to follow suit, and set up a Funeral and Friendly Society in September 1920. Both of these were independent of the Pumpherston Oil Company; for once, none of the managers was on the committee and it was organised and managed by the men themselves.

These two organisations were yearly societies, and so they operated as both a health insurance and savings scheme. Both were in difficulties by the mid 1920s, almost certainly as a result of the economic slump of the period. Strikes and reduced wages meant that for many workers, inessential expenses such as Friendly Society contributions had to be cut, just at a time when they might have been most useful. A prolonged strike in the shale industry took place in 1925 and had a disastrous effect on the two friendly societies. In that year, the Mining Department Friendly Society failed to pay a dividend, and no more is heard of it; that same year the Works Benefit Society was wound up and its funds disbursed to the members. Small local societies such as these often had a short life. With their limited membership and funds, one major claim or a run on their funds could use up their reserves. There may also have been a general notion that in cases of need, the Company would provide.

COMMUNITY SPIRIT

'There was a community spirit that just doesn't exist now. Everyone was in the same boat. The gaffers, they were a bit better off, better houses and that, but everyone else was the same, no one had a lot more than anybody else. In those days the saying was: If you can't pay for it, don't buy it.'

Nursing Associations

A Nursing Association was a body set up to raise funds to employ a nurse, whose services were made available free to subscribers and their families in need of nursing care. It was a valuable service, but an expensive one for a local community, as the nurse's wages, insurance, medical supplies, housing and transport (usually a bicycle!) had to be paid for.

At a public meeting in the Institute Hall in 1924, a Nursing Association was formed for Pumpherston, Uphall Station and District. William Gray, Works manager, was the first president, and Nurse Elliot was the first nurse employed by the Association – 'a thoroughly trained competent and experienced nurse'. In her first year, she dealt with 193 patients, making 2,660 visits. Funds were raised for her support from Scottish Oils, Pumpherston Recreation Club, and by donations and collections at Pumpherston, Uphall, Roman Camp, Oakbank and Hopetoun works. 'Lady collectors' were employed to collect subscriptions round the doors.

In 1929, Nurse Elliot left and was replaced by Nurse Hall. In 1935, the Association acquired a house for her at 16 Letham Park, one of the three-roomed Scottish Oils Houses. Of her 121 cases in 1934-5, over one fifth related to maternity and child welfare. The nurse was clearly moving towards a more modern district nursing role of welfare rather than simply sick nursing.

The work of the Association continued during the war, and Nurse Hall who died in post, was replaced by Nurse Martin. Contributions were raised from 1d to 2d a week from October 1941. By 1945, a car had been acquired, and Nurse Martin was granted a basic petrol ration, and given free use of the car at any time.

The creation of the Welfare State in 1948 promised care from the cradle to the grave, and virtually ended the need for both friendly societies and nursing associations. Pumpherston and Uphall Station's Nursing Association was wound up.

Throughout its existence, Scottish Oils supported the Association financially, and several of the managers or their wives served as honorary presidents. Did the paternalism of the Company – generous, but perhaps oppressive – inhibit community self-help? The fact that Pumpherston did not set up a nursing association or a viable friendly society until the 1920s, while most communities had done so 20 or 30 years before, might suggest so; but on the other hand, the assurance of help from the Company probably encouraged the setting up of many of Pumpherston's clubs and societies.

Doctors

Doctor John MacLardy and his daughter Dr Nina MacLardy. In later life she became an active and forceful councillor for Pumpherston and Uphall Station.
(J. MacLardy)

Pumpherston Oil Company, in common with most of the large mining concerns, also organised medical provision for its workers and their families, by appointing one of the local GPs as its Works' doctor. The doctor's fees were paid by a deduction of 3d or 4d a week from pay packets, but the doctor's services were then free at the point of need. This system of a Works' doctor did provide the services of a doctor at an affordable rate in the decades before the introduction of the National Health Service, but denied the workforce any choice in their GP

In 1913, this 'doctor question' was the talk of the village, when public opinion, expressed in a series of public meetings, forced Dr Stewart of Uphall, the Works' doctor, to allow the Pumpherston workforce to have Dr MacLardy attend their wives and families if they chose. 'Deafening cheers greeted this announcement' reported the *Courier*.

Dr John MacLardy was the longest serving of the local doctors, having come to Pumpherston in 1902. He was noted for 'his neatness of mind and body, a neatness that amounted almost to meticulousness... The characteristic of neatness and punctuality he had carried into his everyday work, and it had made him an inspiration to all.'

In his later years, Dr John MacLardy was assisted, then succeeded, by his daughter, Nina, who became a GP in the days when much prejudice still existed against women doctors. She had the strength of character to withstand opposition and win over the doubters. Her nephew, Dr Iain MacLardy, was the third generation of the family to practise medicine in Pumpherston.

So a doctor's help was available when required. If however the patient required hospital care, the case was slightly different. The 'voluntary' hospitals such as the Edinburgh Royal Infirmary were supported by voluntary donations, and in order to be admitted, the patient required sponsorship from one of the subscribers such as Pumpherston Oil Company, Scottish Oils, or the churches. Alternatively he could go for free treatment to one of the Poor Law hospitals, maintained by the local Parish Councils – such as Drumshoreland.

It was a haphazard health system that worked to some extent, though there were certainly some who slipped through the net. For the elderly and infirm, there was always the dread of the poorhouse – a much feared and humiliating fate. The coming of the National Health Service in 1948 removed the fear of illness and destitution from the lives of ordinary working people.

The Co-operative

The Co-operative movement was intended to provide far more than just shops – it was a workers' movement formed to combat the evils of capitalism, encourage community involvement, and improve the circumstances of working class people through self-help. Later, commercial considerations came to the fore; but there is no doubt that the Co-ops made a significant improvement to life in many of the shale and coal mining towns and villages of West Lothian in the late nineteenth and twentieth centuries.

West Calder Co-operative Society was formed in 1875 and was the most successful of the local societies. It opened up premises in surrounding villages, including Pumpherston in 1887, and took over several smaller societies.

West Calder Co-operative Society was run by working men and was independent of the oil and coal companies in its area, though clearly its success was dependent on the prosperity of these companies to give their workers enough money to spend in the Co-op.

In 1887, a mere three years after the setting up of the village, West Calder Co-op opened premises in Pumpherston. 'A good van trade was being done prior to this, but after we had a Branch and were able to carry local stocks, the trade rose by leaps and bounds.' The fact that seven vanmen were employed in 1911 shows

Pumpherston branch of the West Calder Co-operative Society, c.1904.
(West Lothian Council Libraries)

the extent of the business and its penetration into the surrounding districts. The Pumpherston Store contained bakery, fleshing, shoemaking and dressmaking departments. The original building soon proved too small and was extended in 1891 and again in 1900 to include grocery, ironmongery and drapery. Added later were departments selling coats, suits and shoes, and butchermeat; and an office and meeting-room. Beside the Heatherwood was the Co-op (later Nobel's) explosives magazine, a fortified hut which held the explosives used in shale mining. In the early years, shale miners were obliged to buy their own gunpowder, and most got it at the Store.

PADDY THE COPE

Patrick Gallagher the Irish labourer was so impressed by Pumpherston Co-op that he went on to found the whole Co-operative movement in Ireland.

'The first Saturday I got my pay I went up to Pumpherston Co-operative Store, a branch of West Calder Store, and asked to become a member. I was told the fee was five pounds, but I was taken in on the instalment plan by paying down one pound. It was as much as I could afford. We did all our dealings in the Store. Some people did not deal in the Store: they said the goods were too dear, and before the first quarter ended, I was sorry for joining. However, when the quarter ended we had one pound dividend coming to us. Sally watched the book carefully and began to compare prices with others. She could not find a single instance of where she paid a halfpenny more in the Store than what other women paid in other shops. If you bought butter you got margarine in other stores; it was down as Danish butter. If you bought butter in the Co-op, you got butter, and so on. Everything of the best was kept in the Store. We soon learned that the people who were members of the Store had money and were saving it....

'When we got the house fixed up we took in lodgers. They paid Sally twelve shillings a week. She bought everything in the Store. The Store paid a dividend of three shillings in the pound on all goods purchased. The second quarter, our five-pound share was paid up by our dividend. The next quarter, Sally left the dividend on deposit and added a few other pounds she had saved to it. This happened every quarter. The Co-operative Store paid five per cent interest on deposits.' (c. 1890)

When the branch first opened it had 486 members and first year sales of £10,870. By 1925, the membership was 1,480, and sales were £92,000. There can scarcely have been a family in Pumpherston which was not registered with the Co-op, and it was also drawing a large amount of trade from surrounding villages.

In 1954 the premises were renovated again, and quoting the managing secretary of the Society, Mr McLellan, the *Courier* reported that 'Self service in the shops may come some day, although he himself thought it was not an adequate substitute for the present system...' However, self-service carried all before it, and Krazy Kuts was opened by the West Calder Society in the old bakery premises in 1967, and became popular immediately. Yet another new venture was the

Original appearance of the Pumpherston branch of the Co-op before the extensions of 1892 and 1900. In 1925 a new and more up-to-date bakery was opened diagonally opposite the existing building, and supplied bread to the Eastern district. It also included ample stable and garage accommodation for the new motor vans. Nowadays it accommodates several small businesses.
(West Lothian Council Libraries)

Pumpherston Discount Centre which specialised in brand-name electrical goods and opened in 1977. 'Our aim is to make Pumpherston the discount centre for the people of Livingston and the surroundings areas.' An early resident of Howden recalls walking down to Krazy Kuts in Pumpherston for her weekly shopping, 'wearing out two prams and a lot of shoe leather.' This was perhaps the last period when Pumpherston could be considered as a centre for Livingston rather than the other way round. In 1968 the Mall shopping centre opened at Craigshill, and the Almondvale shopping centre in three phases between the 1970s and 2000.

West Calder Co-op merged with the West Lothian Co-operative Society (already a union of several smaller societies) in 1979. It in turn became Scotmid in the early 1980s. Scotmid closed its premises in Pumpherston about 1982.

West Calder Co-operative Society was an outstandingly successful example of community self-help, which opened premises in Pumpherston before that village was mature enough to start its own society. In 1889, the Pumpherston members of the West Calder Society asked to be represented on the board, and were henceforth allowed to elect two members of the committee. Board meetings were held in Pumpherston twice a year.

While the Co-op existed, Pumpherston attracted trade from the surrounding villages; but the improvement of bus services meant that Bathgate was easily accessible for everyday messages, and Edinburgh for more expensive purchases. The growth of Livingston shopping centres finally put paid to any chance of Pumpherston's expansion as a shopping centre.

THE CO-OP

'You had everything you needed here in the village. They had a Mutuality Club – you could get goods from elsewhere, from the West Calder Store, and you paid so much a week over twenty weeks. There was vans going all over from the bakery. Dod Davis in the 1950s, he came and took your order and then came back and delivered it. Tom Vickers, he was the butcher for years – where the carry-out is now. There was the bakery, the grocery, the drapery and the butcher's, and upstairs shoes and everything and toys at Christmas. Now all the clothes you can buy in Pumpherston is a pair of tights.'

Other Shops

Before the Co-op, the housewives of Pumpherston, if they were not able to walk to Mid Calder or Uphall, were dependent on the village shops such as they were. In the first year of the village's life, a number of Pumpherston wives opened little shops in their front rooms in order to supplement the household income. Pumpherston Oil Company soon put a stop to that, sending letters to offenders: 'We have to request that you clear out your stock of produce within two weeks… as we can not allow our dwelling-houses to be utilized for any other purpose than they are let for.'

The 'official' store in the village was leased from the Company and the store keeper wanted a monopoly of local trade. This store however was not a 'truck store', at which workers were obliged to shop; they could take their custom where they liked. Despite the presence of the multi-department Co-op, the Company shop, known as the 'Old Store' was able to carry on and survived in the hands of various lessees. It also served as the village post office.

After early restrictions on the keeping of shops in houses were relaxed, several residents kept a little shop in their front room; among others, Nellie Smith and Kate McLay sold sweets, and Tam Mabon kept a general shop.

THE PARAFFIN HOOSE

'The Post Office was in one of the houses in the Auld Raws. There was the wee Paraffin Hoose next door. You got your paraffin at the Post Office.'

With the early presence of the Co-operative Store, few other shopkeepers ventured to open a shop in Pumpherston, although there was a chemist's shop by at least 1909. The present chemist's shop was previously a general store, begun by Michael Wynne in a house in School Row, South Village, and run for twenty years after his death by his widow, Mary. In 1958, the Wynnes built new premises, on the site of a wooden hut which is still remembered by older residents as Jock Armit's sweetie shop.

CHIP-VAN FIRE

Tam Gillespie of Uphall Station had a horse-drawn chip-van. The van had a coal fire in it and on one memorable occasion, the van caught fire while standing at the Pumpherston Farm road end. Davie Peden got a police commendation for getting the terrified horse free of its harness. Stein Ramage, who worked one of the many smallholdings on the outskirts of the village and was a noted piper, also had a chip van in the village with his brother Jock. They would stop at the top of the School Row, where the van was so besieged that he often never made it to his second stop at the Old Rows. As the business progressed, he set up shop where the current fish and chip bar now stands.

A number of Bathgate and Broxburn shops sent vans to Pumpherston. Between the wars, Bobby Dunlop from Uphall Station sold fruit and vegetables from a horse-drawn lorry. McClure the Bathgate clothier sent a man round the doors with a suitcase full of samples. James Gray the Mid Calder baker came round with a van. The swans, Jock and Jenny, which used to nest in the Works pond, would take an occasional walk round the village, and one day they had the misfortune to meet Mr Gray's van. His eyesight was poor and one of the creatures did not survive the encounter.

Today those who do not have access to a car have a very limited choice of shops in the village. The ironic fact is that Pumpherston today is less well provided with shops than at almost any point in its history. Thanks to the prevalence of car ownership and Livingston New Town, many small shops have been forced to close. Larger shops congregate in shopping malls in the centres of population. Nowadays, the non-car-owning Pumpherston housewife has further to travel for most of her shopping than her great-great-grandmother a century ago.

CHAPTER EIGHTEEN

Wartime

THE FIRST SIGN OF the conflict was the black-out which was introduced immediately on the outbreak of war in September 1939, and was strictly enforced by the air raid wardens. Sirens, sandbags, salvage, the Home Guard and rationing were other potent symbols of the war years. Youngsters carried gas masks to school, and air raid shelters were built in strategic places. The danger of living next to the Oil Works, an obvious bomb target, did not cause any great concern. Some troops were billeted at Pumpherston Farm for a time, and several local girls married servicemen.

The dark skies over Pumpherston were lit by the searchlights of the anti-aircraft battery at the Track. In June 1940, a returning German Heinkel bomber jettisoned its bombs over the area, and one fell at Mid Calder. Another fell on the stable block at Howden (now Howden Park Centre), killing Mrs David Fleming

An air raid shelter at Pumpherston Works, encased in SOL bricks and made from a converted oil boiler, fitted inside with benches, c.1940.

(Almond Valley Heritage Centre)

and her grand-daughter Margaret, aged ten. They were the first civilians to be killed on the Scottish mainland, and the only West Lothian civilian casualties of the War.

The civilians of Pumpherston were not idle; a War Work Party was formed who raised funds for the war effort. 'High class concerts' and dances were held and 'comforts' were knitted for the troops. Harry Lauder opened a Sale of Work in May 1944 and returned to give a concert in the Institute Hall in November. Sir Harry was played to the Institute Hall by the Pipe Band, and was reported to be 'in fine form and delighted the audience with some of his old favourites, including 'The End of the Road'... 700 people paid for admission and many more had to be turned away.'

So much money was raised that (with the help of a group of emigrant 'Pumpherstonian' fund-raisers in Detroit) the Pumpherston War Work Party was able to present a Mobile Canteen to the WVS and an ambulance to the Red Cross. Mrs Caldwell, wife of the Works manager, was president of the War Work Party, and Mr William McDowall was its secretary. After the ceremonial handing over of the two vehicles in March 1941, 'the proceedings closed with a march past of the Home Guard, ARP services, VAD, Guides and Scouts... led by the local Pipe Band.' Later in the war, Willie Guy of Broxburn who was serving in the forces in Europe, wrote home: 'A couple of days ago a YMCA canteen pulled in here for us and on a metal plaque affixed to the side of it was inscribed 'Presented by Pumpherston War Work Party, Midlothian, Scotland'.'

In March 1942, the government ordered the setting up of Invasion Committees – oddly at a time when the real threat of invasion had passed. They were to be the focus of resistance should Germany have invaded. In Pumpherston its members were William Gray, Works manager, William McDowall, P. Dudgeon, George Grant, George Wilson and Miss Hutton. These committees proved to serve no useful purpose and were soon wound up.

In June 1942 an Area Fire Guard was formed to watch for incendiary bombs and extinguish the fires they caused. It required 70 civilian volunteers, '30 for the North Village, 30 for the South Village, and 10 for the Small Holdings' – making fourteen full parties. Robert Kennedy was Senior Fire Guard for South Village and John McWilliams for the North Village. Like the Invasion Committee, the Fire Guards were set up too late to be useful and faded out.

By May 1945 the prisoners of war were arriving home, among them Gunner Robert Dick and Driver Thomas Anderson, two Pumpherston brick works' employees who had spent some four years in German POW camps. When the war ended, the Scouts cut a stairway up the side of the bing beside the Track and built a huge bonfire as a spectacular celebration of the peace. The bonfire was lit by Mrs Molloy who had six sons in the forces. One of them, 22 year old seaman-gunner Patrick Molloy, was killed on active service.

CHAPTER NINETEEN

Public Services, Transport and Livingston New Town

PUMPHERSTON OIL COMPANY WAS among the first to install telephones in its works, and so the Oil Works had the telephone number Broxburn 1. The first public telephone in Pumpherston was not installed in the Post Office until 1923, and it was after the Second World War before a telephone became an affordable luxury for the majority of Pumpherston folk.

The Co-operative Society was also a pioneer of telephones and electricity in its stores as early as the 1890s, and the oil companies were much quicker than the coal companies to install electricity in mines and works. For the villagers, however, electric light was late in coming. Scottish Oils installed the necessary equipment in 1923, and the Institute Hall and Public School no doubt thankfully threw out their paraffin oil lamps. Gas was not brought into the village until 1988.

The first public transport serving Pumpherston was set up by SMT (the Scottish Motor Traction Company) in 1910, connecting Pumpherston with Mid

Part of the South Village, 2001.
(Helen Scott)

BLOWING A FUSE

'The Works generated its own electricity and the tenants got their electricity from the Works, for 6d a week – with one rule: no electric cookers or washing machines as they tended to blow the fuses. The fuses were in outdoor fuse boxes, sealed, and if you blew the fuses you had to get a Works electrician to come out to it, and very likely a ticking off. In the early evening, the electricity was switched from the Works to the National Grid. When the lights dimmed and came up again, that was the changeover. Sometimes swans flying into the cables caused power cuts.'

Calder, East Calder and Edinburgh. The journey took 65 minutes and there were five buses (or 'cars' as they were known) on weekdays and ten at weekends.

Bus services expanded between the wars and Pumpherston was well provided with services to Edinburgh, Bathgate, Broxburn and the other surrounding towns and villages. In 1956 the minister reported that Pumpherston had over 50 bus services daily, and that a large number of local men and women commuted by bus and train to Edinburgh. Today there is a continuing need for good public transport services in the village in view of the low level of car ownership, and in fact some 70 buses a day pass through the village, connecting it to Broxburn, Livingston, St John's Hospital, Bathgate, East Calder, and Edinburgh. Furthermore, just under a mile away is Uphall Station on the Bathgate to Edinburgh railway line.

A useful survey of Pumpherston, carried out by local schoolchildren in 1969, tells us that 184 Pumpherston residents worked in the village, and 410 worked elsewhere – a complete change from the days when only an oil company or Co-op worker could find accommodation in the village. In addition, the schoolchildren's survey provides the less useful but more interesting information that the village contained 91 dogs, 42 cats, 56 budgies, two tortoises, eight guinea pigs and a parrot! Of the 519 households, 174 had a car (33.5%); and almost 93% had a television.

Car ownership has increased over the last thirty years and now stands at 51.8%, but this is well below the West Lothian average of 70%. Only Addiewell, Blackburn and Blackridge had a lower level of car ownership in 1991. Despite (or because of) its proximity to Livingston, Pumpherston is in the lower half of West Lothian communities when various prosperity indicators are examined. Whereas Pumpherston started out with a young population, nowadays 18.4% of its population are pensioners. Its proportion of owner occupiers is lower than the West Lothian average. And 3.3% of Pumpherston's households are lone parent households – exactly the same as a century ago, though the condition was caused then by bereavement rather than by separation or divorce as today. All these are 1991 Census figures.

Today (2001), Pumpherston's population is 1,341, living in 561 households –

an average of 2.4 persons per house – a far cry from the average of 5.7 per household in 1891. Pumpherston has come a long way since then, but still has some way to go before matching the prosperity of some other communities in the county. Unless the Drumshoreland proposals go ahead, the population is expected to remain static, and no new housing is planned for the immediate future within the village boundaries. The farmland to the south and east of the village is protected as part of the Livingston Countryside Belt.

Livingston New Town

For all its paternalism, the Pumpherston Oil Company thwarted the establishment of alternative industries in the village. Like all company villages, Pumpherston found it impossible to diversify or to attract any other industry or jobs. For 80 years the workers were almost all accommodated in Company housing and had no freedom to move to another employer without losing their home, so other industries did not come. Then with the advent of council housing, employees gained independence from the Company, and the freedom to seek employment elsewhere. Other, non-Company workers came into the village, bringing a valuable infusion of new blood. There was no sudden influx of new industries, as all the heavy industries of the central belt were in decline, but a new phenomenon was on the horizon.

In 1961, the Secretary of State for Scotland announced that Livingston would be the site of Scotland's fourth new town. The location was chosen for several reasons; it had good road and rail communications to Glasgow and Edinburgh; it was at a convenient distance from Glasgow for overspill residents to go back and visit their families; it was one of the few sites in central Scotland where old mine-workings did not prevent major development; and it was hoped that the unemployment caused by the decline of West Lothian's shale and coal mining industries would be relieved by a new town acting as a focus for economic growth.

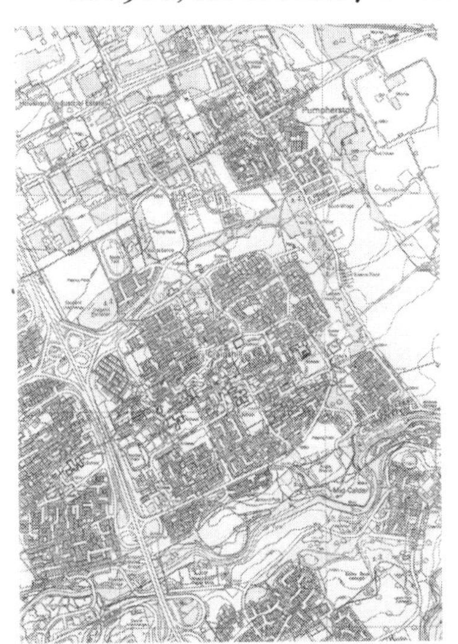

Pumpherston today, on the edge of Livingston New Town.
(Ordnance Survey Landplan, 1998, 1 : 10 000)

Since Pumpherston was to be cheek by jowl with the new town, the news of the designation was received by locals with a mixture of anticipation and dismay. There would be many new jobs, but at what cost to their quality of life? However, with the imminent death of the shale oil industry, there was a cautious welcome both for the new town, and for the other major new development at that time – the British Motor Corporation plant at Bathgate (later British Leyland).

The initial growth of the new town was gradual, giving the local people time to adjust to the new presence in their midst. Craigshill – very close to Pumpherston – was the first district of the new town to be developed, and Houstoun was the first industrial estate.

In 1964, the firm Ready-Mix (Edinburgh) Ltd set up a concrete-making plant in Drumshoreland Road, which was officially opened by Brigadier Purches, general manager of the New Town Corporation. This was the first firm independent of the oil companies ever to set up in the village. Thereafter, however, the proximity of Livingston New Town made it difficult to attract new businesses to Pumpherston. With all the incentives to settle in a new town, what was there to attract a firm to Pumpherston? In 1966, Cameron Iron Works opened on new town land that had formerly been part of Nettlehill and Milkhouses farms and by the end of the decade it employed nearly 2,000.

As part of local government re-organisation in 1975, Pumpherston, Kirknewton and the Calders were transferred from Midlothian to West Lothian. The perception was that these villages economically and socially looked more to the West Lothian towns of Broxburn, Livingston and Bathgate, than to Midlothian. Pumpherston shared in the economic trough of the 1980s, when many large West Lothian industries – British Leyland, Plessey, Golden Wonder – closed down, and unemployment soared. However, a 1981 study of deprivation in West Lothian showed that Pumpherston had escaped the worst effects which were so evident in Addiewell, Stoneyburn, Blackburn and other small communities. To some extent it was cushioned by the close proximity of jobs and facilities in Livingston.

Another boon to Pumpherston in the 1980s was the re-opening of the Bathgate to Edinburgh railway line. Commuting to a job in Edinburgh became much easier, and the communities close to Uphall, Livingston and Bathgate railway stations came to the attention of private house builders. McLean Homes built Beechwood at Uphall Station as a direct result of the railway re-opening. In 1990, Pumpherston's first private housing estate was built – Heatherwood Park – and there is currently the possibility of several hundred new houses at Drumshoreland.

LIVINGSTON NEW TOWN?

'Everyone of my generation hates Livingston New Town. It took in 32 farms. Craigs Farm belonged to Tommy Henderson and his family and they'd worked very hard. They had to leave the farm – compulsory purchase. We used to walk over there and feed the ducks on the pond. It was a sin to build on it.'

The economic benefit of Livingston New Town in terms of employment is obvious. The more intangible effects of the New Town on Pumpherston are harder to measure. The shops, facilities and jobs it offers are welcomed, but older people generally regret its coming, citing an increase in crime and vandalism. Certainly the old Pumpherston was a tight-knit community, sharing the same employer, work, housing and social life. Older residents often recall that everybody knew everybody in the old days. Nowadays, with a population almost exactly the same as it was in 1891, people may not even know those in the same street. Many residents spend their working lives away from the village, and the number of households is more than double what it was. People generally live a more private, less communal life. There is no doubt that the rapid changes of the last 40 years have brought some loss of community spirit, many more incomers, and an increase in rowdyism and vandalism.

However, incomers and change are not necessarily evils, and it would be difficult to find any community in which vandalism has not increased. There have been perhaps as many benefits as losses: greater tolerance and open-mindedness, more opportunities for young people rather than the assumption that the Works was the only future for the boys and domestic service or shop-work for the girls. And after all, Pumpherston too was a new town in its time – an urban community mushrooming on the green fields of the 1880s, a youthful population in which almost everyone was an incomer. From that disparate population, Pumpherston managed to forge a strong sense of community largely free of sectarian divisions, and to retain it for over a century, even in the face of encroachment from its giant new neighbour. Perhaps Livingston has a thing or two to learn from Pumpherston.

PART THREE

Social Life, Sports and People

by

JAMES O'HAGAN

CHAPTER TWENTY

Recreation

Recreation Committee

IN THE EARLY YEARS of the village's existence, the Recreation Committee was at the centre of local affairs. It met under the chairmanship of the Works manager of the time and consisted of the secretaries of the leading organisations, the district councillor and union secretary Joe Heaney and permanent treasurer Jimmy Dunlop, the well-known *Courier* correspondent. All the essential financial business of the village was transacted there and meetings were held once or twice annually to consider the financial assistance to be given to the various clubs and associations. Any village group producing a financial statement for the year automatically received a grant according to its needs and circumstances. Requests for help were seldom refused and there was some consternation when the thriving Junior Football Club, made more or less independent by the proceeds from their new clubhouse and bar, decided they did not need the money and did not submit their balance sheet. The footballers were the first to show the way and as clubs and organisations became more self-reliant and developed ways to raise money for themselves, the Recreation Committee found itself redundant.

The Institute Hall

In the early days, a room in the public school was used as a reading room for the villagers, but space there was soon inadequate and the impressive Works Institute was opened in the summer of 1891. It consisted then of a large meeting room and a reading and amusement room, plus bowls pavilion and a house for the hall-keeper. The role of the Oil Company as guardian of public morals was emphasised at the opening when Mr William Fraser spoke of the workmen now being able to enjoy innocent games or an hour's quiet reading, and hoped that it would put a stop to a game very common on the road sides which brought discredit to the whole community – tossing pennies. It was, though, another fifty years and more before the old gambling game died a natural death in the shale mining communities. The only beneficiaries of the pastime were the village youngsters who were paid to perch in the nearby trees and signal the approach of unduly conscientious policemen.

In spite of that popular and illegal counter-attraction, the reading room and

The Institute Hall, 2001
(Helen Scott)

amusement room were crowded every night in those early years. Life, though, was not all billiards and dominoes. Among the clubs making use of the hall were the Pumpherston Orchestral Band and the Pumpherston Musical Society while the Pumpherston Literary Society was formed a few years later.

Demand was such that the building was extended in 1907. A large main hall and stage was added and a library with a new reading room provided. The roof of the main hall is of particular interest with its heavy arched beams made from laminated timbers bolted together, and many concert artistes have paid tribute to the fine acoustic qualities of the building.

Gala Days

Until the middle of last century the Gala Day and the annual seaside excursion known as the 'Store Trip' were the highlights of the year for the unsophisticated village juveniles. As late as the end of the 1950s, attendance at the Gala Day was more or less obligatory for all youngsters of school age. Initially the event was part-funded by a voluntary yearly donation of one shilling from the Oil Works employees. The Works management gave the Gala their full support and as a consequence there were few workmen who cared to dissociate themselves from the event. This meant there was ample help available for the setting up of the running and jumping areas, transport of benches from the Institute Hall, distribution of the 'bags', and litter collection at the end of the day. The office staff recorded the

Pumpherston Gala Day 1911, looking south. The procession is coming up to the Store Corner. Some of the children have 'tinnies' on a string round their necks, ready to be filled with milk or juice. *(Almond Valley Heritage Centre)*

results and looked after the important task of distributing the prize money. At its peak the title Gala Day became a misnomer and a full week was given over to subsidiary events such as the long and high jumps, the 'marathon', five-a-side football and netball with occasional fund-raising adult events such as pram races and sponsored walks. In the 1920s the committee catered for up to 1,300 children.

There were always two bands in attendance at the Gala with the West Calder or Broxburn Brass Bands being considered, by the children, much inferior to the village Pipe Band. There was particular derision for the man with the big drum when it was discovered he had only one drumstick.

The 1960s brought changes as the older children became more worldly-wise and wearied of the long trek round the village behind the musicians. They learned to arrive at the ground along with the van bringing the 'bags', and marshalling the crowds became a major operation.

There was never any shortage of volunteers on the actual day of the event, but as the influence of the oil companies waned and the Works management became less involved, it was difficult to find people prepared to tackle the onerous and time-consuming tasks of raising funds, distributing tickets and organising the Gala Night dance. Committee members tended to drift away as their children outgrew the entertainment and there were years when lack of adult interest meant there was no Gala Day.

There were some resurgences. In the 1970s a committee with George Hamilton in the chair took a fresh look at the set-up and, among other changes, ini-

William Mackay (right), Pumpherston Headmaster, with Works manager Fraser Cook at a Gala Day in the late 1950s.
(J. O'Hagan)

tiated the annual election of a Gala Day queen. Members like Irene Clusker who served for twenty years were the exception rather than the rule, however, and as support fell away the event lapsed again. A last revival came in the 1990s under Ian Armit when a strong team including secretary Evelyn Rogers and local piper Archie McIntosh raised several thousand pounds to give the Gala a new lease of life. Sadly if predictably village support was not all it might have been, the effort lost momentum and at the time of writing there have been four consecutive years without Galas.

Oatridge Races

For adult villagers, or at any rate the males amongst them, the equivalent of the youngsters' Gala Day was the annual Oatridge point-to-point race meeting. On that day Pumpherston was a half-deserted village as men and youths trekked the handful of miles to the course via Uphall, the various hostelries there and the fields beyond. For most of the horse racing years, Pumpherston had no licensed premises and the presence of a refreshment tent was as great an attraction as the remote possibility of showing a profit on the day's transactions. The standard of the horse-racing was high; the standard of some of the bookmaking rather lower.

Youth Organisations

The Pumpherston Scout Troop first came into being in 1922. In 1923 its registration was sanctioned by Headquarters and, since Pumpherston was then in Midlothian, it became the 36th Midlothian Troop. John Nicol was appointed Assistant Scoutmaster and at the same time, a Cub pack of twenty boys was formed under Assistant Cub Master William Anderson. The Girl Guides were organised by Captain Mary Petty and all units were regular attenders at local church parades in those early years.

By 1926 Scout Master Murray was presiding over more than 40 Scouts and Rovers and the Midlothian Scout rally that summer was led by pipers from the Pumpherston Troop. The Cub Pack was in the hands of Miss Katy Begbie and

Miss Alice Linden who were to be leaders for another 25 years and who were both decorated for their services to the movement. Popular local man Davie Armstrong became Scout Master in the early 1930s and was in turn succeeded by John Harvey. The organisations continued throughout the war although meetings were held in the Institute Hall for the rather macabre reason that the Scout Hut had been commandeered as a prospective mortuary.

John Harvey, a baker with the local Co-op, was still in charge by the time the Scouts returned to a fairly casual existence in the hut off Drumshoreland Road. Activities consisted mainly of week-end camps held at Jock Ross's farm, Contentibus, at Dedridge and there was a spell at the tennis pavilion at Howden House after the camping equipment had been stolen. John Harvey was immortalised by the more musical campers who frequently serenaded him with original words to the tune of 'London's Burning'. 'Skipper Harvey, Skipper Harvey; We're starving, we're starving; Give us food, give us food, We'll die of starvation.' In fact, meals were substantial although they never varied much from fried bread with beans, bacon and sausages. At least in retrospect, those holidays by the banks of the Linhouse were idyllic. There was complete freedom with unexplored wildernesses, interminable football games, camp-fires and the sound of distant trains through the night.

The more forward-looking Jock Newton took over the troop in the late forties with Rab Young and Will Newton as his deputies. 'Skipper' was dropped in favour of 'Scouter'; and uniform, almost non-existent during the war years, became more

Pumpherston Rovers and Scouts before a trip to Holland in 1947 or '48. The Scout leader (centre) is Jock Newton.
(J. O'Hagan)

or less obligatory with the tartan neckerchief giving way to a more modern blue and white one. Jim Clark and Nola Buchanan were the popular leaders of the Cub Scouts. Camping was still high on the scouting agenda and there were moves further afield, to Eyemouth once for a fortnight, and later still to the Continent.

When the Rover Crew was formed Jock Newton became Group Scout Master, leaving Rab Young as Scout Leader. Bobby Dudgeon presided over the strong Rover group with Dobbin and Tammy Miller, Jim and John Clark, Geordie Wilson, the Horsburghs and many others. Family connections were a feature of the Rovers who had a rejuvenating effect on the village, starting a football team, running the Gala for some years and organising Harmony Nights and Country Dancing classes which were extremely popular.

In the 1950s the Scout troop was at its strongest with first Bill Fairley, then John Clark taking over the post of Scout leader which was becoming ever more demanding. By the mid-1970s the popularity of Scouting was declining both nationally and in the village and John Clark resigned after being associated with the movement for more than forty years, half of that time as Scout leader. There was a revival, engineered by George Hamilton, at the end of the 1970s but it was short-lived and the village Scouts and Cub Scouts passed out of existence.

A branch of the Boys' Brigade was formed in 1891 under the command of an ex-Northumberland Fusiliers sergeant, Andrew Gordon. They were supported by Peter McLagan who promised a set of drums to the Band being formed. Perhaps fortunately for the villagers this failed to materialise. The Brigade never enjoyed the popularity that the Scouts did. There were occasional appearances in Gala Processions and there was still a company under Captain George Ferme in the late 1950s before it too disappeared.

Several attempts to start Youth Clubs, notably a gymnastic one by headmaster Mackay in the early 1950s and another by Mrs Mulholland and interested adults a few years later, had early success but faded out of existence.

Pipe Band

Pumpherston Pipe Band was formed in 1908 and the Oil Company produced the money needed initially for uniforms and instruments. On this occasion the company directors led by William Fraser, donated the necessary funds and the band played in the Fraser tartan.

By the start of 1909 the membership was eighteen of whom twelve were learners. Pipe Major Angus Livingston held practice classes twice a week. It cost one shilling to join the Band and there was a threepenny weekly subscription. When another three sets of pipes, practice chanters and music were purchased, the bill was less than twenty pounds.

There was great enthusiasm when the pipers made their first march through

Pumpherston that Hogmanay. The celebratory circuit of the village became a tradition which was followed whenever the band scored a success in major competition. There is a story that they returned so late on one occasion that, out of consideration for those villagers who were asleep, they removed their boots before beginning to play.

Perhaps their most memorable moment came at Cowal Highland Gathering in 1946 when they won the Sir Harry Lauder Shield. It was the last time that Sir Harry presented the trophy in person to the winning band.

Other successes included the Argyle Trophy at New Meadowbank in 1945, the Anderson Shield under Pipe Major Matt Moir at Dunblane Highland Gathering and the 4th Grade World Championship in 1963 under Pipe Major Alex Cupples. This last was followed by the Scottish Championship in Princes Street Gardens two months later. Another noted Pipe Major was Charlie Manderson who served with the Royal Scots through the First World War. His composition 'Drumshoreland Muir' deserves a wider audience.

It would greatly upset these veterans to know that Pumpherston Pipe Band disbanded in the 1970s. Falling numbers and the birth of a new band in Livingston managed what two world wars could not.

Pumpherston Pipe Band in front of the old Golf Clubhouse, c.1930
(West Lothian Council Libraries)

CHAPTER TWENTY-ONE

Sports

Recreation Ground and Athletics

OUTDOOR INTERESTS WERE CENTRED on the recreation ground and the bowling green. These were both built in 1891, just a few years after the first houses and at the same time as the village hall. At the turn of the nineteenth century running, cycling, and jumping were every bit as popular as football. In 1891, for instance, several hundred people turned out to watch the Bathgate Cycling Club's first race and by 1893 the Pumpherston playing field had been enclosed by a cycling and running track. The ground, officially Recreation Park, was and still is known more familiarly as 'The Track'.

The first 'gymnastic games' took place in the village Sports Park in August 1886. The main event in a varied programme was a 130 yard open handicap which required nine first round heats. There were, among other activities, a tug-of-war (Pumpherston beaten by Broxburn), a triple jump (won at the equivalent of twelve metres), quoiting and 'cutting the goose'. The games were held annually thereafter and the first meeting at the new Recreation Park was in June of 1891 when the 120 yard sprint was won by a Pumpherston man called, appropriately, J. Hare.

Over the turn of the century the emphasis was on sporting events along the lines of the Highland Games. By the 1930s, though, interest in running and cycling had declined and football was the more popular attraction. Open meetings where a few of the races were set apart for village entrants were gradually replaced by more local gatherings with one or two open invitation events. Apart from venues such as West Calder where annual Highland Games were kept alive for another couple of decades, the athletics meetings held in most villages turned into or merged with Gala Days for the benefit of local youngsters.

Although track and field events never regained the prominent place they once had in local circles, there have always been village athletes good enough to perform with distinction in national competitions. Works foreman joiner, Davie Blain took part in the heavy events in Highland Games in the 1940s. Davie made himself a caber which no-one else in the village could lift, and, to every-one's amusement, padlocked it to the golf-course fence for safety. Johnny Loch and William Lowe were respected competitors on the professional racing circuit, while more recently John Allan has had considerable success on the track. He will be remembered, particularly by his backers, for winning two distance races on successive afternoons at one of the New Year meetings at Meadowbank in the 1990s. The only local cyclist who made any mark was Ricky Lumsden, a member of the three-man team that won the Scottish Hill Climb back in the 1960s.

Football

The first village football team appears to have played in 1891 when the *West Lothian Courier* carried an account of a Pumpherston eleven beating Edinburgh Teachers 2-0. The newspaper, in prophetic mood, declared that with a few more matches, the newcomers would take a prominent place in local football. This proved accurate enough after a fashion, for a month later they faced Linlithgow in the East Scotland Shield and lost 9-1.

The Pumpherston Juniors football team which lost 2-0 to Shotts Bon-Accord in the Scottish Junior Cup final in 1958.
Back (l-r): R. Muir, Jimmy Murphy, Ian Bennett, W. Farmer, Vince Halpin, Davie Johnstone. Front: John White, Joe Young, A. Craven, R. McKay, J. Berry. Trainer, J. Cummings.
(West Lothian Courier)

From then on the village produced a succession of football teams in various minor grades. Generally they were competent enough without being world-beaters. Few survived for more than a year or two, although Pumpherston Rangers had a long run in the 1920s and won the Midlothian league and a trophy or two in that time. Between the wars notable local players were Hector Wilkie and William Dornan of Hibernian and Tommy 'Boxo' Rogers who played for Heart of Midlothian.

More recently Archie Murphy was a great favourite at Alloa and Walter McWilliams was a Hibernian signing. Archie was described by the famous Scottish internationalist John White as 'the best wing-half I ever played beside', while Walter turned out in the Hibs 'Famous Five' forward line. Archie's daughter Elaine was a full internationalist with the Scotland women's eleven, and Walter's son Derek played for several senior teams including Dunfermline, Dundee and Falkirk. In a related field, Con Duggan, a contemporary of Walter and Archie, became President of the Scottish Schools' Football Association and is currently Education Officer of the S.F.A.

In 1954 a group of enthusiasts led by Will Watt (the 'big drummer' in the Pipe Band) did considerable volunteer work on 'The Track', bringing it up to the standard required for Junior football, and Pumpherston Juniors was born. John Caldwell accepted the post of Honorary President, Davie Miller was the first secretary and a fifteen-strong committee worked behind the scenes. A voluntary contribution of one penny a week from all of the work force at the oil plant provided the necessary financial backing for the team and a fine building with changing rooms and showers downstairs and the club premises above was designed and seen to completion by Ian Angus of the Works drawing office staff.

Although the Club's registered colours are black and gold, the team initially played in a maroon strip, a set of Heart of Midlothian jerseys gifted by the famous Scottish international player, Tommy Walker from Livingston Village. The Juniors gathered a large and faithful support without ever quite achieving the success they deserved and the rather disappointing high point of their existence was reaching the Scottish Junior Cup final in 1958, when they were defeated by Shotts Bon-Accord at Hampden.

Captain Vince Halpin and local favourite Jimmy Murphy (a younger brother of Archie's) were two of their most popular and longest serving players. Davie Blain, the Highland games 'Heavy' athlete, was trainer for several years. Davie never ever knew his own strength and injured players preferred to jump up and run rather than have Davie administer his brand of muscular first aid.

Other village players were popular members of neighbouring Junior clubs, particularly Bert Rogers at Ormiston, Tom Hill with Bo'ness and Jim Lowe at Broxburn. Jim's younger brother Drew was signed with Heart of Midlothian for a time.

Support for the Juniors dwindled and there was a long decline before the Club wound up about 1972. In the late 1980s a new group of energetic enthusiasts with Tom Peden playing the leading part resuscitated the team, and an application to re-join the Junior League was accepted in the 1990 season. Since then a hard-working committee has ensured that the club is again firmly established in the Junior football world.

With the birth of the Juniors, the local young people lost the use of their traditional home ground and the notorious 'Iodine Park' was constructed alongside the Track. The playing surface was coarse spent shale and minor injuries were so frequent that local doctor Nina MacLardy provided the Rover Scouts with a monthly supply of bandages, plasters and iodine. The pitch was eventually brought to an acceptable standard and a succession of teams, mostly run by the Scouts and Rover Scouts, played in the minor Lothian leagues. There were few successes – although the Scout team's victory in the Billy Ritchie Trophy in the Midlothian Youth League in the 1960s is still remembered – but there have been years of priceless enjoyment for the young villagers.

Bowling Club

Traditionally the Bowling Club originated in 1887, but the first report of its activities comes from 1891 when the Institute Hall and the Green were opened. The

Jim McGinty, winner of the Linlithgowshire Bowling Association County Singles Championships in 1981.
(J. O'Hagan)

game must have been new to the ladies at the ceremony, who were given a chance to play before the Company chairman's wife, Mrs Wood, rolled a bowl along the green to open it and a match between Mr Bryson's team and Mr Caldwell's team took place. The village quadrille band helped to provide music for dancing and the *Courier* reported memorably that when the band left, it did so 'to the evident chagrin of the devotees of Terpsichore'.

There were some desirable trophies presented that season with David Burton receiving a gold badge and cup while Charles Thomson won prize bowls and Mr Bennet won a pair of silver mounted bowls. The trophies, unfortunately, have been lost.

For the first half of the last century the bowling green was a focal point for the villagers and particularly the older inhabitants. After the war there was a gradual realisation that bowls was no longer simply an old man's game and while the atmosphere was still relaxed, the Club became more purposeful and ambitious. It chose the same way forward as the footballers and with the acquisition of a bar, presided over by Freddy Jamieson, there was an immediate improvement in the Club's financial standing, eventual expansion of the premises and an increase in social activities. Little chagrin has been displayed by the devotees of Terpsichore since then.

All of the district trophies, Livingston, Carmichael, Hopetoun, Meikle, Rosebery, Fraser, MacLardy, have come to the club at various times, with a highlight being the first capture of the Colville Jubilee Cup by a rink skipped by Davie Armstrong in 1956. Perhaps the most impressive spell was the memorable summer of 1985 when George McCulloch jnr, Ronnie Wardrop and George Rarity took the SBA Inter-district Trophy. That year the Club were also successful in the Ladies' LBA Singles and Ladies' LBA Pairs. The following season they won the Carmichael Trophy, LBA Ladies' Champion of Champions and Midlothian County Championship.

Pumpherston, though, are still seeking that elusive first Scottish title. They came closest as recently as 1999 when David Anderson jnr and Derek Wardrop won the Scottish District title and moved into the National stages. Eventually they were defeated by Prestwick in the final but they had silver medals to console them for a commendable performance.

Not all of the Club's famous victories occurred on the green. In 1982 there was a national success and television appearances for Con Duggan, Jim Robb and Stewart MacFarlane when they won the Scotsport Sports Quiz.

The honours board of Club Champions is graced by the names of some fine players with no one ever having had a monopoly of success. An individual performance worth noting, though, was Jim McGinty's victory in the LBA County Singles Championship in 1981, the only time the Club has been successful in this prestigious competition. The representatives for the sponsors, a well-known whisky retailer, worked hard to make sure their product had a prominent place in all the photographs and breathed a sigh of relief when the hullabaloo was over and they ushered the victor into the Clubhouse asking him if he felt like a small refreshment. 'Thank you very much,' said Jim, 'I'll have a dark rum.'

The Cricket Club

The Cricket Club's appearance on the village scene was marked by a game played against a Broxburn eleven in 1885. Pumpherston won narrowly with the kind of score that was all too familiar in later years: 34 runs to Broxburn's 32. A month later they scored eleven all out in a home game against Fauldhouse Victoria. Things improved the following year under an expanded committee and the captaincy of William Gowans. By the early 1900s they had won the Linlithgowshire Secondary League playing against Bathgate, Broxburn and Armadale seconds.

Former site of the Cricket Club's pitch, looking east towards the Oil Works. The old cricket pavilion can be seen in the centre distance.
(J. O'Hagan)

At this time the team, with assets estimated at three pounds and five shillings, was comfortable financially but lived dangerously at the wicket. Finding a safe piece of level ground for a pitch proved difficult and various sites were tried before the Club made their home at Letham Park. The pavilion was opened there in 1923. Top bowler Alex Skene became captain, there were sound bats in Simpson, Stoddart and Porteous and by the 1920s the Club were running a second eleven of their own and playing fifteen games each season.

The War stopped play in 1939, but moves for a revival began soon after it ended. It took some time to find a suitable field and lay a square, and for a year or two there were practices on the edge of the football field between pitch and running track. At that time the newly re-formed Golf Club was losing some of its land to the expanding Detergent Plant at Scottish Oils. The Cricket Club volunteers lifted the fairway turf and Charlie Carr and Willie Madill laid the square in Magazine Park beside the Heatherwood, roughly where the fifteenth green of the new golf course is. It was quite an achievement with the only mechanical help being a couple of wheelbarrows. Part of Magazine Park (so-called because it was next to the Explosives Magazine) had been divided into allotments during the 'Dig for Victory' war years but those had largely disappeared by the time the cricketers took over. A metal fence kept the cattle off the square and had to be removed before each match.

The Cricket Club was in a strong position by the end of the 1940s. The ubiquitous Joe Heaney was installed as secretary and George McWilliams, a captain of the Club before the war, became president. During all the post-war years, the captain and guiding light (albeit a rather unsteady light at times) was James 'Peely' Armstrong, while Robert Frame and Andrew Nicholson were enthusiastic senior assistants. For the most part though cricket had become a minority interest in the shale villages and there was no longer an official league. Practice nights in summer were popular but younger members did not care much for spending long Saturdays in the field and there was often difficulty in producing teams for friendly fixtures against such opponents as Merlin, Bangour Hospital, Leith Franklin, Holy Cross, Rosebank and Muirhouses.

The Marshall brothers, Willie and Alex, were regulars, with Alex, the wicketkeeper and successor to Peely as captain, being noted for stopping the ball with whatever part of his body he could manage to put in the way. John Allan, capable of hitting sixes to every part of the ground, was the main batsman. Of the younger players, Tammie Miller and Ian Loch were promising fast bowlers and among the others, the names of Gordon Hill, Bobby Allan, Jimmy Caldwell, Stephen Heaney and Tom Harper stand out. Later on outsiders arrived and the team lost a little of its village flavour but for a few halcyon years, locals would wander over and sit spectating in the sun beside the pavilion on a Saturday.

Sadly, interest in the club waned. With the field being used infrequently, communications with the farmer leasing the land broke down and the club was evicted, ostensibly for not paying the 50 shillings rent money. For a time the square survived, a pathetic fenced-off area in a sea of wheat, but it disappeared eventually. The old pavilion, reinforced by bricks and mortar, still stands on the original site and is now the home of the Pigeon Club.

Pigeon Club

In the early days the pigeon fanciers in Pumpherston were a collection of individuals rather than an organised club. The fashion was to have a do-it-yourself doocot (some of quite remarkable construction) in the back garden or in the Heatherwood and most enthusiasts belonged to the neighbouring East Calder Club. Dod Gardner, Neil Carr, Sandy Gowans, Tam Patterson and George Wilson were well-known names, but Jack MacDonald (the long-term treasurer of the Junior Football Club) was by far the most successful practitioner.

By 1961 there were enough people racing pigeons to justify the formation of a club and the Cricket Club premises conveniently fell empty the next year to provide the new organisation with a headquarters. There were early difficulties which included accidentally burning down the wooden hut. Some fanciers amalgamated with their Livingston neighbours to form the Pumpherston and District Club off the Mid Calder road, but the Premier Pigeon Club, under President Sandy Gowans, continues to thrive on the old cricket site where the clubhouse is now built of brick.

The Golf Club

The Golfing Annual of 1896-97 lists Pumpherston Golf Club as being instituted in the Summer of 1895 and there is a record of land being rented for a course at Harrysmuir the following year. The Oil Company provided the necessary financial backing and the deed for the lease was signed by James Caldwell who was Pumpherston Oil Company manager then. It was the start of a long association between the Caldwell family and the Pumpherston Golf Club – an association that remained unbroken until John Caldwell, managing director of Scottish Oils, died in 1970.

The first course at Harrysmuir, where Houstoun Industrial Estate is now, was on a flat, featureless piece of ground and had nine holes ranging in length from 200 to 330 yards. The Club's stay there was brief and in 1908, James Caldwell officially opened a new nine-hole course on waste land beside the Oil Works. The Club, with a membership of around fifty, played there for more than twenty years and it was during this spell that it joined the Linlithgowshire Golf Association. In 1931 there was another move and a third nine-hole course was constructed on the west side of the main road through the village. The Letham houses occupy the

Phil Smith, John McLean and David Anderson snr, Pumpherston Golf Club, with the *Courier* Trophy – 1979.
(J. O'Hagan)

site now. Again the course was short, level and, by modern standards, unimaginative in lay-out.

The mid-1930s, though, were successful years competitively speaking and a strong team of W. Loch, T. Mabon, A. Smith and R. Rae won the much-prized *Courier* Trophy for the first time while W. Nathaniel travelled to the County Boys' Championship at Bathgate and brought back the Walker Cup. The Letham course closed during the Second World War amid what must surely have been far-fetched worries that it offered a possible landing place for invading enemy aeroplanes.

It was to be another 43 years before Pumpherston won the *Courier* Trophy again and the silverware was accepted on behalf of the Club by the same R. Rae who played in the first victory. Robert was the Club's longest serving member until his death in 1999. He was an office worker at the Oil Works, a popular self-effacing character and a valuable and almost permanent committee member. His golf, in keeping with his character, was consistent and dependable.

Of the other players in that notable team, Albert Smith was chief cashier at the Oil Works. Bert was very tall with a rather agricultural swing, but though his action looked awkward it was efficient enough and he was always liable to conjure up a remarkable shot or two when he was in trouble. Tam Mabon was a jack-of-all-trades who provided haircuts, and did upholstery work, watch repairs and the like in a hut at the foot of the New Rows before moving to a permanent shop still in existence near the Cawburn Inn. Sometime Club Champion and course record holder at Ingliston before it became home to the Royal Highland Show, Tom died in 1993. William (Wull) Loch senior drove tankers for the Oil Company and was, by common consent, the best local golfer of that era. Not particularly long but very accurate off the tee, he had an excellent swing and a dependable putting stroke. Wull was well-known for his habit of walking to church in Mid Calder carrying his brightly coloured golf umbrella and scandalising the more soberly attired citizens. He did not readily give up the search for a lost ball, hunting assiduously while muttering grimly to anyone in earshot, 'It's no the value o' the thing – I'd just like anither skelp at the wee bugger.'

When the war ended the Club negotiated a lease with Walter Dandie, who had taken over Pumpherston Farm from Danny Miller the previous year, and the course and clubhouse moved back to the site occupied in 1908. The Oil Company organised the dismantling, transport and re-erection of what became known as the 'Old Hut' and laid on water and electricity. A new lounge was built in 1971 and there were further changes in the 1980s when the last part of the wooden building that had served for almost a century was eventually demolished.

There is no record of a Club Championship at any time before the 1939-45 war and there was none for the first years of the Club's re-formation. Indeed there were no regular competitions at all until Jack Crombie (who worked in the laboratory at Pumpherston and was the perpetually unsuccessful Tory candidate in the local elections) donated an annual prize for junior members. Golfers of a nostalgic turn of mind may like to know that the lucky winners received a Dunlop 65, a Blue Flash and a North British Twin Dot.

The first Club Championship was in 1960 and was won fittingly by President John Ritchie. It was his friend, Angus Sinclair, a whisky salesman, who initiated the event by providing a suitable trophy. John, the foreman engineer at the Oil Works, saw the Club through some of its more difficult times before he left the district in the 1960s. He died in December 1999.

Members brought up in an age of manicured greens, well-cropped fairways and rough no more than ankle high might like to be reminded of the conditions under which their predecessors played. The fairways were cut four times a year (when the Club could afford the ten shilling charge) and the job was done by Bert Dandie (then at Primary school) with his father's tractor and hay-mower. The Oil Works lent out labourer Willie Madill one afternoon a week to do the greenkeeping and this included hacking down the rough with a scythe. One of the Club's problems was that Walter Dandie kept four working horses as well as assorted cattle, sheep and occasionally a bull on the course. For a long time the greens had to be protected by barbed wire fences but this was a minor inconvenience compared with the need to take evasive action when a companion's ball came to rest on part of the course that had already been used by one of the larger farm animals. Another natural hazard was subsidence, caused by old mine workings or the collapse of an underground pipe. There have been several incidents over the years – 'a bit alarming' commented one insouciant member, 'but not life-threatening unless you were on the spot at the time'.

In the late 1950s the Club had a bit of good luck when Jimmy Smith, a relative newcomer to the village, came over to an Annual General Meeting out of curiosity and accepted the position of treasurer. He admitted long afterwards that he did so on compassionate grounds when he saw how disorganised the committee was. Over the next thirty years Jimmy worked miracles on the most slender resources and the Club would not have survived without him. He oversaw the installation of the bar, and, single-handed, put Pumpherston Golf Club on a sound financial footing by the time he retired.

The most difficult moment for the Club came when Walter Dandie and his eldest son Dave of Pumpherston Farm decided to take the golf course land back into the farm. At a subsequent general meeting it was clear that the younger members were determined the golfers would not suffer the same fate as the cricketers had done a few years previously. Led by such forceful characters as Jim Lowe, Stephen Heaney, Ian Loch and Frank Conway, the members explored every possibility of staying in existence, whether on the same site or elsewhere.

After protracted negotiations and the involvement of both the County Council and the local MP, the District Council agreed to give financial backing. The lease of the land was put on a firmer footing and for the first time the Club was in a position to make future plans with a degree of security.

The Club had still to husband its income prudently and greenkeepers remained part-time while the bar was staffed by volunteers. Under Jimmy Smith's guidance, though, there was an unprecedented spell of success on the playing, social and, particularly, the financial fronts. The old hut was proving inadequate for an ambitious membership and Tom Jenkins and George Cleland organised a band of helpers to construct a new bar and lounge, a project that had previously been beyond the Club's wildest dreams. This in turn led to an expansion of social events, permanent bar staff and the employment of a full-time greenkeeper.

Golfing honours came along at an unprecedented rate, most sparked off by a trio of scratch golfers, John McLean, Davie Anderson, and Phil Smith, all of whom were county champions. Phil died tragically in a climbing accident on Ben Nevis, but John and Davie are still going strong and are a match for anyone in the district. The Club also struck a rapport with two very different top professionals: British Ryder Cup captain and Bathgate-born Eric Brown and USA Open Champion, Steve Elkington, both of whom accepted life memberships.

The Club has always been well served by its Committee members, and two in particular who remember the years when committee work meant appearing in working clothes, are still serving. Former secretary and current co-ordinator Jim Lamond has done much outside his official remit during his long association with the Club; and Jim Lowe, equally hard-working and long-serving, was the architect and prime mover behind the four holes which gave the Club its first decent lay-out a few years ago and led indirectly to the new course.

The characters employed as greenkeepers always added much to the free-and-easy atmosphere in the Pumpherston clubhouse. In the 1950s the extrovert Wull

Gowans, a fine singer much in demand at Burns Suppers, was at a loss for words just once in his life. A visiting golfer with eye-catching flowing locks was asked by Wull, 'Whae the hell gave ye that bluidy haircut?' The stranger, perhaps slightly upset after an indifferent round, replied, 'The same boy that cut thae bluidy greens.'

In the 1980s it was the often irascible Tam Murphy who reigned supreme over the fairways and he took great pride in his many mentions in the *Scotsman* newspaper's 'Country Diary' column, written by club member, Jim O'Hagan. Tam was seated outside the Clubhouse one scorching day as two strangers struggled up the last hole to discover to their dismay that the bar had just shut. 'Oh, sir,' pleaded one, 'You couldnae get a couple o' thirsty men two pints o' beer, could you?' Tam shook his head regretfully. 'Naw,' he said, then allowing exactly the right amount of time for the bad news to register, added, 'but I could get three thirsty men three pints o' beer.'

CHAPTER TWENTY-TWO

Village Notables

Some Village Characters

CHARLIE MCKAY, THE LITTLE tramp who wandered tirelessly round the Calders area for the last half of last century, lived secretly in the New Rows for a time. Nellie Smith, like several of the village women, ran a sweet shop from her living room and Charlie, terrified of being called up to serve in the Army, stayed there in hiding and took exercise outside at night. Later he spent his life wandering the roads dressed in a succession of too-large Air Force greatcoats. When the current garment wore out he walked to Newcastle where the Salvation Army gave him a replacement, insisting, much to his disgust, that he had to have a bath first. On one occasion, he made the long trek to find that the Hostel had been burned down – surely the ultimate in wasted journeys. He spent his last days in the Old People's Home at Limefield where the matron wryly observed that although he ran away frequently, he always returned by tea-time.

Daddy Duffin, presumably an Irish immigrant, lived in the middle row of the

Wee Charlie McKay the tramp in his Salvation Army greatcoat.
(J. O'Hagan)

New Rows and often sat outside the house playing the fiddle. In those days the road (of stones and hard-packed dirt) between the houses was busy with girls (and often housewives) playing skipping ropes and beds, while gangs of boys indulged in interminable games of three-and-in, leave-o, kick the can and one-a-side headers with the coal-house doors for goals. Daddy spent his days pushing a wheelbarrow between his home and the allotment garden he tended on the other side of the small heath (the Heatherwood) which extended to the village doorsteps. He was always accompanied by a small terrier and never tired of asking the local children if they knew the difference between an elephant and a pofor. Much to his delight they always asked the obvious question.

The Oil Company generally enlisted fairly hardworking employees but somewhere along the way Harry Donnelly managed to slip through the net. He left school during the depths of the Depression, and was unable to get a job. Thereafter he never got into the habit of working. A good-natured man, Harry was to be found perpetually loafing about the Store corner. On one occasion he lasted no more than an hour and a half in a gainful employment before returning home. Questioned about the reason for such a short stay, he declared he felt guilty about doing two horses out of a job.

Jock Armit, the owner of the village sweet shop, a wooden hut sited where the chemist's shop is today, was an important person in the lives of the younger villagers. His domino-playing cronies used his shop as a meeting-place and the counter was the domino table. It was usually necessary for customers to wait until a chalk was finished before they were served. As far back as the 1940s the hut boasted the first one-armed bandit in the village. Jock owned an ancient black Labrador called Bess. Near the end of its life its hindquarters were paralysed and Jock took it to the shop every day holding its rear end in the air while the dog padded away on its front legs. He would probably have been arrested in the current climate but the dog seemed contented enough.

In those days no rural scene was complete without a village policeman and Pumpherston was favoured with one who was absolutely typical of the breed. P.C. Thomas Dudgeon, known as 'Coughin' Tam' because of his non-stop smoking, was a large man who was certainly not built for speed, particularly when attired in his stiff uniform and regulation boots. Life would have been a sinecure for him had it not been for the younger villagers setting fires, playing in forbidden areas and devising new ways to steal apples from the fruiterers' carts. In those days the penalty for such crimes was a clip on the ear but Tam was seldom able to get close enough for the ends of justice to be served.

There were many others residents worth mentioning. George Cleland the popular foreman of the Wax Sheds at the Oil Works was a noted painter of the local countryside. Jimmy Wilson, whose nocturnal habits led to the nickname of 'The Ghost' was often accompanied by a pet pigeon which perched on his shoulder. Hall-keeper Jock Kilpatrick found a fine use for the bowling pavilion when

Pumpherston men in front of the Institute Hall. Left to right: Hughie O'Hagan, unknown, Jock Pennykid, Cock Hunter, Jock Armit (with Bess)
(William Gold)

he used it to cure his home-grown tobacco during the years of shortage. The local butcher, long before there was any demand for equal opportunities for women, was Mrs Leslie who would obligingly skin, gut and prepare poached rabbits at her kitchen sink. Geordie McWilliams, a president and captain of the Cricket eleven was a self-educated man and fine musician who gave violin lessons. And, of course, every one knew red-headed Alex Smith, not as sharp mentally as his fellows and peering through thick spectacles as he shuffled his way through the village in a small contented world of his own.

More Recent Villagers and Activities

Oil Works foreman joiner, Davie Blain, already mentioned as an athlete and football trainer, was well-known in other fields. His feats of strength were renowned; he was Sunday School superintendent and his first action on becoming trainer to the Junior Football Club was to ban swearing in the dressing-room; and he was a keen local historian who laid the first steps for an account of the village. His work led a group of senior citizens organised by Victor Armstrong to attempt to produce a short booklet on Pumpherston's history. Sadly Vic died before it could be printed, but his booklet provided one of the starting points for this book.

There has always been a strong musical tradition in the village from the first years of the Institute Hall when a Pumpherston Orchestral Band, a Madrigal Society and a Musical Society were all formed. Later the fashion was for less formal musical groups and concert parties such as Mrs Crombie's 'Sunbeam Follies',

Dr Nina MacLardy's revue artistes and Gerry McLauchlin's 'Seven Belles' who raised much money for charities over the years. From the 1940s to the 1970s, few weeks passed without a concert of some kind in the Institute Hall.

Davie Blain, 1903–1987, active in many aspects of village life all his days.
(Marjorie Lamond)

One of the 'Sunbeam Follies' singers was Bobby MacKerracher who won the Britain Bing Crosby competition before an audience of 4,000 at Edinburgh Palais in 1945, but the village boasted other fine vocalists in Jock Ferguson and Johnny Murphy, while Jimmy Armstrong had an enviable reputation as a violinist. Bobby MacKerracher was the grandfather of Kerry McGregor whose television appearances in 1997 excited much admiration well beyond the village. Kerry was lead singer with the group Do-re-mi when 'Yodel in the Canyon of Love' just failed to become Britain's entry in the 1997 Eurovision Song Contest.

The area as a whole was well served by dance bands composed of local personnel. Joe Wood, a fine bass player and occasional singer, started the Woodchoppers Trio in the late 1950s with Fred Mulholland on piano and drummer Bill Sharp. Later he formed the equally popular Meltones and both groups are still remembered with affection. Stephen Heaney, the youngest of Councillor Joe Heaney's sons, was also a versatile singer and played double-bass with Willie McNeil's Downbeats, the resident band at the Edinburgh Locarno ballroom in the 1960s and early 1970s.

In the 1970s younger musicians emerged and Alan Robertson, Jackie Ness and Pete Irvine combined in the Almond Valley Trio. For a time Alan and Pete were the driving force behind the Pumpherston Jazz Club in the Seven Oaks on Saturday afternoons. While still a schoolboy, Alan Robertson formed a seven-piece jazz band in which he played trombone. Later based in London, he developed into a talented arranger and accompanist, appeared regularly on television, had a pop hit record 'Speak to me Clarissa' under the pseudonym of Alan Trajan, but sadly, died in 2000.

There was a Burns Club, the Pumpherston 'Bonnie Doon' Burns Club, established as far back as 1926. Mr Aitchison proposed the Immortal Memory to an audience of 100 at their first anniversary Supper and for some months there were regular meetings although the Club does not appear to have lasted long. The Golf Club restarted the Burns Supper tradition in 1973 when Ian Angus brought in artistes like orator Bob Brown, singers Jim Speirs and Gibby Sutherland, reciter Will Kirk and Scottish champion fiddler Celia McIntyre to perform at consistently over-subscribed events. An occasional speaker was former villager Bill McDowall. Bill, currently writing about wine and whisky for various Continental publications, was noted for his appearance on television as captain of the Edinburgh University team in 'University Challenge'.

The local branch of the WRI was established in 1935 with the wives of the Scottish Oils hierarchy playing a strong supporting role. Village names which occur again and again are those of Mrs W. McDowall, Mrs W. Hunter, Mrs A. Linden and Mrs J. Roberts. The ladies involved spent countless woman-hours raising money for charities, collecting for Cancer Research, supporting other organisations and entertaining senior citizens.

This tradition of providing help and entertainment for the older villagers became a labour of love and has been the concern of a band of volunteers for nearly half a century now. Present secretary of the group, Nan Lawrie has been involved for 35 years organising the trips and treats that mean so much to the senior members of the community.

Sporting personalities ranged from quoiting expert Joseph Sneddon in the 1890s to boxer and hockey international Wattie Gardner in the 1930s, and to motorcycle champion Robert Miller in the 1950s. Wattie was a goalkeeper, no job for the faint-hearted, and it was widely claimed that he dealt with awkward high balls by stopping them with his head. In later years he cheerfully acknowledged that the habit might have had something to do with his consequent unworried outlook on life.

Robert (Rab) Miller, nicknamed 'The Goon' because of his various crazy juvenile exploits. While still at school he managed to fit a motor-cycle engine to his bicycle, and he was also noted for descending the local shale bings on his push-bike. These exploits tended to overshadow his undoubted skill on the track, but he was an accomplished rider and won many trophies, including the Scottish Speed Championship in two successive years.
(J. O'Hagan)

David McKenzie was a recent Scottish International basketball player both at School and Senior levels. Six feet tall before he was fourteen, Davie received a bit of a shock when he met a visiting team of fourteen year olds from the United States and discovered some of them were a good head taller than he was.

There were many other minority interest in the village at various times including clubs for anglers, karate enthusiasts and chess players, as well as residents noted for activities as diverse as greyhound racing, dancing and showing cage birds. Much material from early years was either unrecorded or has been lost and it is inevitable that some organisations and many individuals worthy of note have been neglected.

The Heaneys

One family which can never be overlooked in the history of Pumpherston, was the Heaneys. No one did more for Pumpherston than Joe Heaney, although his daughter, Elsie, must have come close. As one newspaper said at the time of his death in March 1967, 'The names Joe Heaney and Pumpherston are synonymous.'

Joe was a member of the Labour Party all his life and typical of the best of the working-class people who formed the backbone of Labour supporters. He began his career of public service on the board of West Calder Co-operative Society, and by the 1930s he was secretary of the village branch of the Shale Miners' and Oilworkers' Union. He became an executive member, then full-time secretary, until the closure of the shale oil industry brought the winding up of the Union. In 1952 he was elected local county councillor, and as chairman of the Landward Housing Committee he had every right to feel proud of his contribution

Councillor Joe Heaney meets the Queen.
(Elizabeth Hepburn)

to Midlothian's, and particularly Pumpherston's, housing record in the difficult post-war years. His final satisfaction came with the building of the 63 houses on the old Heatherwood site – a development which now bears his name. His work helped to stem the exodus of young people from the village, and contributed to Pumpherston's continued viability as a community.

His funeral was one of the largest ever seen in the district and people like Alex Eadie MP, Tam Dalyell MP, chief constable William Merrilees and a host of other dignitaries were among those paying their last respects. There was unanimous and unstinting praise for Joe's work and for the integrity and sincerity he showed throughout his life.

Joe's daughter Elsie followed in her father's footsteps by joining the Labour Party on her 18th birthday. She became a District Councillor in 1969 and a County Councillor for Joe's old constituency two years later. In all she served for over 22 years, actually two years longer than Joe, and she did so with all her father's dedication and clear-sightedness. There were no 'surgeries' in those days but Elsie was so often out and about in the village (later, villages) that surgeries were unnecessary and, anyway, her door was always open to anyone who needed help. There was a great sense of loss at the next elections when there was no Elsie Hamilton standing by the door checking off the voters, every one of whom she knew, as they passed in and out. She could tell the result of every election long before the votes were counted. She was described by her fellows as one of the finest councillors they had the privilege of knowing, a tribute she really deserved.

PART FOUR

Greening Over

by

KNEALE JOHNSON

CHAPTER TWENTY-THREE

After the Closure

THE DETERGENT WORKS AT Pumpherston closed in 1993. For 110 years the Works had been part of Pumpherston; indeed the village owed its very existence to the shale oil industry. Mindful of the long association between the shale oil industry and the local community, BP took immediate steps to counteract some of the consequences of closure. The company Powles Hunt bought part of the detergents business at the site, and was granted a lease of part of the remaining office block on favourable terms. The company went from strength to strength, and at the end of the year 2000, was able to move into a purpose-built office and manufacturing premises near the M8 motorway at Coatbridge.

To help create new jobs to replace those lost at Pumpherston, BP entered into an agreement with the West Lothian Enterprise Trust to fund an additional local economic development worker for two years. At the end of that period the project had clearly been so worthwhile that a further year's funding was provided.

BP then began to consider the future use of the site. It was decided early on that it must be used for the benefit of the local community. Much work remained to be done in removing redundant plant and examining the environmental condition of the site before a suitable use could be decided, but the concept of a 'beneficial' use was to be a guiding principle in all the work that followed.

The environmental legacy

The environmental condition of the site was the major concern. In the century and more since the Pumpherston Oil Works was built, there had been a sea change in attitudes to, and understanding of, the environment, and environmental awareness was an area where BP was determined to be a leader. But on a site with a long history such as Pumpherston, there were likely to be a large number of unknown factors and unrecorded pollutants.

In the early years of the Works, the practices adopted for waste disposal would be inconceivable today. But 50 to 100 years ago, companies all over the country would discard waste materials in open pits or down disused mines, or bury them and then build anew on top.

It was known that tars – the waste by-product of the tar distillation and refining processes – had been stored in specially constructed ponds, and that over the years waste chemicals had been buried on site. Although there were operational records going back many years, it was not known whether they were complete.

The Soap Works in the 1970s. Filling bottles with Sainsbury's washing-up liquid. *(West Lothian Council Libraries)*

To make matters even more complex, the use of the site had changed several times: mining followed by refining, then detergent manufacture and oil refining, and finally detergent manufacture alone. Oil and water do not mix, but detergents are designed to help them do so. What effect might the combination of these substances have had upon the soil at the Pumpherston site and the underlying watercourses? In particular, had any of these potential contaminants shown any inclination to spread beyond the site and how could this be prevented in the future?

To get to the bottom of all of this, a specialist team was brought in by BP to investigate the site and advise what should be done. The team included specialist environmental and soil experts from BP's international research centre, as well as engineering and project management staff with experience of similar projects elsewhere. This team had worldwide experience in tackling difficult contaminated land problems and had already been involved with some smaller shale oil related activities in West Lothian.

Although the best possible people were brought in, all projects of this kind are different. It was to prove necessary to develop a number of special techniques to solve Pumpherston's problems; some of these new ideas would be applied elsewhere after being proved at Pumpherston. Such is the nature of technological progress, but it is pleasing to reflect that the spirit of innovation that led to the site's early success was still alive over a hundred years later when the process of 'greening over' began.

The investigation begins

With so much history to investigate, it was natural that the first reaction was 'Where do we start?' There were several avenues to be explored before beginning work on the site, and this preliminary work would decide what work was necessary, and where.

The first approach was to search what might be called the formal memory of the site itself – in other words, the records, maps, plans and drawings resulting from all the years of technical activity. The quantity of material was daunting, but was it complete? Activities that we would regard as significant in today's environmental climate might simply have gone unrecorded decades ago. These formal records gave a good general guide to how the site had been used in the past and they helped with targeting the engineering investigations, but it was soon clear that they could not be relied upon for a definitive view of the site's condition.

The next avenue was to search the informal memory of the site – the recollections of people who had worked there. Many such people, including several general managers of the site, were tracked down and interviewed. Sometimes they were contacted again as work proceeded, to check a fact or a hypothesis that was emerging from the desk study. Such personal contacts were invaluable and gave the whole exercise a strong feeling of reality as opposed to a piece of purely academic research.

A valuable source of information was aerial photographs. Sets of photographs were available at various times going back to the Second World War. Careful analysis of these using modern techniques helped to pinpoint where things were on the site at that time, and what changes had occurred over the fifty years since.

While these studies were taking place, consideration was being given to what exactly was required from the site investigation. It needed to find out the level of contamination on the site so that an appropriate clean-up strategy could be devised. It also had to find out whether any contaminants were leaking from the site, and if so how this could be stopped. So the investigation was to extend well beyond the site itself.

Boreholes and trial pits

The preliminary work enabled detailed planning of the physical investigation to take place. It was clear that since the main concerns were soil contamination and associated contamination of ground water, a good deal of soil and water sampling would be needed. A comprehensive programme of sinking boreholes, digging trial pits and analysing the resulting samples was begun. The location of the various sampling points was chosen to give good general coverage of the site and surrounding areas, but also to focus on the most likely areas of contamination as revealed by study of the records.

In all 34 boreholes were sunk (seventeen of them as deep as 47 metres), and 77 trial pits were excavated. From all of these as well as a further twelve sampling points, samples were collected and a variety of on-site and laboratory tests were conducted.

Results of the investigation

The investigation concluded that there were only two classes of contamination – detergents and hydrocarbons – that needed treatment or management. There was no serious contamination by heavy metals.

The most encouraging conclusion was that migration of contaminants from the site did not pose a high risk to either the local population or the environment. Most of the hydrocarbon contaminants were in a form where they were essentially immobile, so their effects were contained within the site boundary. There was evidence of detergent contamination in groundwater and surface waters but this was not harmful to humans, although they would have an effect on aquatic organisms.

The site investigation pointed towards two main tasks – to clean up the water leaving the site in order to protect local water courses from detergents; and to deal with the hydrocarbon contamination of the site itself.

CHAPTER TWENTY-FOUR

A New Beginning

SOON AFTER THE CLOSURE of the plant in mid 1993, demolition of the manufacturing plant and most of the associated buildings began. By the end of the year the site looked quite different. The high columns and mass of pipe-work associated with detergent manufacture had gone, but a few key installations remained, at least for the time being. Demolition was not just for cosmetic reasons; it allowed BP to carry out its site investigation work more easily. Two warehouses were retained in case they were required during the clean-up or as part of the future use of the site, but in fact they were demolished towards the end of the clean-up phase.

The office block at the north-west corner of the site was also retained. Built in the 1970s, it proved a valuable resource during the intensive work on the site in the late 1990s. In the year 2000 it was named James Young House in honour of the founder of the Scottish shale oil industry and was placed on the market as a free-standing office block in what was becoming an attractive new environment.

Perhaps the most important part of the plant to escape the demolition gang was the effluent treatment plant. It was built to process contaminated waters before their discharge from the site, and its value during the site clean-up was obvious from the start. Later, once the contamination levels had been reduced and a self-managing system of water discharge handling was in place, it too was demolished.

So as the twentieth century drew to a close, only James Young House was left standing. Its name provides a potent reminder of the site's past and the early growth of an industry of huge importance.

The clean-up starts

With demolition completed and the site investigation complete, it was time to start the clean-up in earnest. No decision had yet been made about the future use of the site, but certain measures were clearly required whatever the site's eventual role was to be.

The first task was to ensure that no contaminated water was leaving the site and reaching the external environment. The site investigation had confirmed that because the site was on top of a hill which sloped away to the south and the east, waters were leaving the site only in these directions. This applied not only to surface water, which could pick up contaminants lying on the site or in the very top layers of soil, but also to groundwater. Contamination could result from rainwater soaking down through several layers of soil into the groundwater beneath.

One of the reed beds which treat contaminated water by trapping impurities in the root systems. Looking south across the Almond Valley.
(Robertson Partnership)

It could then flow into local watercourses.

Deep drainage ditches were constructed along the southern and eastern boundaries of the site. These intercepted the groundwater flow before it left the site and enabled the water to be collected for treatment. At first the water was pumped back up to the old effluent treatment plant. Much of the resulting water was then discharged to the public sewer under normal licensing arrangements. But a longer term solution was required, so that in future years, drainage from the site could be self managed and would not be dependent upon pumps and treatment.

A system of reed beds was therefore constructed below the site and near the Bank Burn. Reed beds are essentially large flooded fields densely planted with reeds. They are carefully constructed with clay linings so that no leakage occurs. The water enters at one end and slowly progresses through the reeds. The root systems of the reeds break down the contaminants in the water using natural processes until, by the time the water reaches the end of the system, the water is essentially uncontaminated.

The reeds used in such a system have to be carefully chosen for the range of contaminants they will encounter, and the design of the beds is a highly specialised exercise. At Pumpherston considerable laboratory testing was undertaken to decide the most suitable type of reeds. The reed beds are huge; two separate and parallel first stage beds, each over 100 metres long, discharge into a roughly circular final stage pond where further reeds treat the water one last time before it is discharged into the Bank Burn, which flows into the River Almond.

The performance of the reed beds proved to be excellent and by 1999, aided by a gradual reduction in contamination levels in the water feeding into them, they could be relied upon to clean up the groundwater flows from the site on their own. The old effluent treatment plant was demolished and these silent natural workers now clean the waters to the demanding standards set by the Scottish Environmental Protection Agency. Virtually no human intervention is required except to keep an occasional check on the quality of the water.

The reeds in the purification beds are Phragmites australis. They are native to our island but have no common English name, and they are now well established and tower over the occasional interested countryside walker. Sedge warblers and reed buntings are taking an interest in the area, while moorhens have already raised families. Whitethroats, blackcaps and garden warblers can be found on the old railway track alongside, and swallows and sand martins forage above the water. The bottom pond is home to a thriving colony of sticklebacks, and as non-flowering plants and small invertebrates increase, the beds will be a haven for more and more of the larger birds and mammals.

Perhaps of greater importance than the role of a specialised and increasingly scarce habitat is the part the reed-beds and golf-course play in linking Calder Woods and Almondell Country Park with the Scottish Woodland Trust property at Drumshoreland. The establishment of such corridors of land is essential for the movement and spread of all our wildlife. Every one of our lowland mammals and most of our plants and inland birds can be found somewhere along this stretch of territory and it will become increasingly important as the years pass.

The Bings

Although not part of the BP refurbishment, the removal of the shale bings is worthy of notice. Around the Pumpherston Works were four bings: to the north, Pumpherston Bing, to the east, Clapperton Bing, to the south, another unnamed bing, all of which were the spent shale from the Pumpherston Works. To the north-east is Roman Camp Bing, the product of the Roman Camp Oil Works. These pink mountains of spent shale have been gradually eaten away for road bottoming and brick-making. Pumpherston Bing is gone, as is Clapperton Bing that filled the triangle of land which was once Pumpherston's common; and the bing immediately to the south of the Works. Extraction of the shale from the Roman Camp Bing is likely to continue for some years to come, but once work is finished, the remains will be landscaped.

Tars

The most challenging part of the clean-up were the tars remaining on the site. They were a by-product of the processes at the Works over the years, and these useless heavy deposits were simply consigned to man-made ponds or lagoons to get them out of the way. In the past, some of these ponds had been partially stabilised by filling them with rubble, but even those would require to be checked to ensure their physical stability, and capped to enable their sites to be re-used.

It was decided early on that simply removing the tars from the Pumpherston site and burying them elsewhere in a landfill site was not a desirable solution. In environmental terms it would achieve nothing, and the removal of the tars to another site would be hazardous. It was therefore decided either that the tars would have to be destroyed away from the site, or that they would have to be dealt with on the site itself.

At first efforts were concentrated upon off-site destruction of the tars. The most attractive solution seemed to be burning the tars in cement kilns so that the very high temperatures would destroy the hazardous components. An added benefit was that during the process the heating value of the tars would be released and could be used. If such a solution could be made to work it was felt that the hazardous transportation of the tars to the cement kilns could be justified.

Despite a great deal of technical work and negotiations with the cement industry, these attempts eventually failed. There were concerns about emissions from the cement works, and the most suitable kiln was so far away that transportation costs and environmental concerns were too great.

While these negotiations continued, research had been proceeding into biological processes which might break down the tars. It was found that the tars to be dealt with fell broadly into two chemical categories; and one of them responded to a new technique of biological treatment, which had been developed in conjunction with a Welsh company, Celtic Technologies. The technique involved a form of landfarming – not in itself a new idea – but finding a way to apply it to the Pumpherston tars was a major technical and economic breakthrough.

After extensive laboratory trials, then field trials, this process was implemented in 1997. The process is known as landfarming because the equipment used is mostly standard farm equipment. This helps to keep costs down. The tars are placed in beds 100 metres long and 15 metres wide, layered with soil, inert material and chemicals. The beds are turned over by ploughing frequently, sometimes daily, to ensure that the oxygen which is critical to the biological process is constantly replenished.

During the process at Pumpherston, eight of these beds were created, and it was found that within three months the mixture of tars and soil had become clean enough to allow it to be used as topsoil in the redevelopment of the site. This was approximately a quarter of the time that similar processes had taken in the past, and this rapid clean-up is one of the major technical achievements of the Pumpherston restoration.

The whole process, which occupied some four acres of land, was repeated twice more, making three times in all. By this method 12,000 tonnes of tars were converted into usable topsoil in less than a year.

There remained some 10,000 tonnes of tars which did not respond to this type of treatment. They were treated on the site by creating specially lined repositories into which the tars were placed. The tars were then stabilised by mixing them in the repository with inert shale and cement. The resultant massive concrete block is below ground level and is physically and chemically stable. After covering the top with a layer of clay, then topsoil, the area was able to be used for redevelopment.

Finding the best use for the site

Right from the start, much thought had been given to the best long-term use for the site. It had to be something which would benefit the local community, but what? Housing development was rejected early on because of the history of contamination on the site; and simply keeping the site as some form of wilderness area was rejected for fear it might become a neglected wasteland.

A shortlist of potential uses was drawn up which included re-development for industry or some form of leisure amenity. Although continuing to use the site for industry would in some ways have been appropriate, the chances of attracting a new industry were not good, given the many incentives in nearby Livingston. So the choice came down to a leisure amenity for the local community.

There was one obvious candidate; Pumpherston Golf Club was next door and had close historical connections with the site and the Works. Their existing course was nine holes, and the redeveloped area would enable an additional nine holes to be built. The only other serious leisure possibility for the site was to make it a country park, but the preference within the local community and the local authority was for the golf course option. West Lothian Council agreed to contribute 40 acres of their land, and this gave enough space to create an extension to the Pumpherston Golf Club course, bringing it up to a full eighteen holes.

Greening over – the golf course project

Detailed feasibility and design studies were undertaken by BP – still the site owners – and the Golf Club, and outline planning permission for the golf course extension was obtained. This allowed Pumpherston Golf Club to apply for lottery funding, and in 1998 a lottery award of £785,411 was confirmed – the biggest award to any golf club in Scotland. The lottery board was particularly impressed that the course was being developed on industrial land that might otherwise have been left derelict.

The golf course was designed by Graeme Webster of the firm, Glen Andrews. Soon construction of the new holes was under way and by the end of 1999 the broad shape of the new course was clear. Its setting and atmosphere make it easy to forget its industrial past. Careful landscaping and tree planting, as well as the retention of existing trees, have produced an attractive course. Below the flat area where the industrial plant stood, the land drops away, providing interesting hilly holes with picturesque and challenging water hazards.

Golf Club captain Alan Docharty welcomes the changes, which will expand the course from 25 acres to 150 acres. 'The old clubhouse will be turned into greenkeeping facilities. The new one on Drumshoreland Road has panoramic views of the Pentlands and will offer improved social and administrative facilities, allowing the Club to open its doors to more Junior and Lady members. For the

Pumpherston from the air looking east, 13th September 1999. The Store Corner is in the centre foreground, and the South Village on the far right. The dark area at the top is the site of the Oil Works, over which the golf course has been extended. The triangle forming 'Pumpherston's part of Drumshoreland Muir' can still be seen.
(John Rae Studios, Larbert)

first time a professional will be employed, and it is hoped to encourage the local community to come and try the sport. None of this would have been possible without the assistance of Sport Scotland (Lottery Award), BP, West Lothian Council, the Royal and Ancient Golf Club of St Andrews, and not least the Club members.'

So the wheel has turned full circle. Pumpherston, a green and pleasant place until the village and Works were created by the shale oil industry, has been through its own industrial revolution. It developed in response to the industrialisation of the Victorian era and it provided a livelihood and a social framework for generations of people.

Changing patterns of work and travel mean that no longer do people need to have their workplace on their doorstep, but a community still needs its social meeting places. What might have been a blight on the village, has been transformed into an asset. Where shale oil was once refined and detergents made, there is once again a 'dear green place' where golf, the game that Scotland gave the world, will be played for years to come.

CHAPTER TWENTY-FIVE

Past and Future

DURING ITS FIRST HALF-CENTURY, the village was run by very few men – the managers of the oil company who controlled most aspects of life – housing, leisure, employment – but who were unelected and unaccountable. It's generally agreed that they were well intentioned, but their primary consideration was not the welfare of the inhabitants, but the maximisation of profits. Paternalism was all very well, but profit came first.

Despite these constraints, the Pumpherston Oil Company meant well by Pumpherston, and intended it to be a model village of its kind. The houses were of a higher standard than most contemporary company housing, and were upgraded several times in the following decades; and within twenty years or so of its founding, the village was remarkably well provided with leisure facilities.

The proximity of the housing to the mines and Works was preferred by both the Company and the workers in order to reduce travelling time and costs; but mines, works, railways and bings in such close proximity led to what would today be unacceptably high levels of pollution. Fumes, dirt, smells, specks of ash on washing lines; polluted rivers and burns – and in windy weather, fine dust from the bings coating everything to leeward – these were the difficulties which faced residents of early mining villages and particularly the women who toiled to keep their families and homes clean and healthy.

The level of supervision – not to say interference – by the Company in the lives of its employees both inside and outside of working hours would be impossible to imagine today. Yet despite these drawbacks, there is little evidence in the memories of older people today of any dislike for the old Pumpherston. They grew up within the system. It was a shared way of life – most of the residents shared the same employer, the same work, the same houses; they lived in close

PATERNALISM

'The Company was everything. The men used to tip their hat to the managers. The manager was the Lord Almighty. If you'd grown up with it, you weren't so aware of it. You didn't think to question it. But it was more than paternalism: they dictated. The manager had his "informer", who slipped down under cover of darkness and told the manager what was going on. Well, that was what people suspected. The manager knew what was going on, and if he didn't like it, you were called in and told off.'

proximity to one another, with shared sanitation, back greens and laundry facilities. It was a more sociable, less private way of life than we are used to today, and it formed strong community bonds, support for one another in times of hardship, and much pride in the community.

PALISADE MENTALITY

'When a woman was sick, the neighbours rallied round; one would make a pot of soup, another came in and said, 'I'll do the carpets', and hang them over the rope on the back green. Whereas now, especially when the houses were sold, not so much among the locals, but the first thing the incomers did was throw up a six foot fence. The palisade mentality – you never, ever got that then. Everybody knew each other.'

In 1883, Pumpherston was an agricultural estate with a population numbered in dozens. A couple of years later it had mines, Works, miners' rows and a population of several hundreds. It was not unique, this new village, for all over Scotland, companies built works on greenfield sites and new villages to accommodate their workers. In West Lothian alone in the half-century from 1850 to 1900, there were many new villages; among them, East Benhar, Westerton, Uphall Station, Philpstoun, Woodend, Addiewell, Stoneyburn, Mossend, Gavieside, Oakbank, Roman Camp. Of these company villages, built by and dependent on a coal or oil company, fewer that half have survived. The others rose and fell with the companies which built them. When the company closed its local mines or works, the villages decayed or were abandoned.

If, for some reason, the Pumpherston Oil Company or Scottish Oils had closed the Pumpherston mines and Works at some point in its first half century, the village would have died. It would have become a ghost village, the fate that befell more than half a dozen other West Lothian communities. Pumpherston had no independent existence outside of the Oil Company; its employment, its housing and most of its social life depended on the Company. It was fortunate in that the Works on which it depended survived longer than most, but it was the council houses that gave Pumpherston some existence independent of the oil companies and enabled it to survive.

When closure did come to the Pumpherston Works, in stages between 1962 and 1993, the village did not die. It survived, for it was no longer dependent on one company; it had a life and momentum of its own. Having survived thus far, Pumpherston will not be swallowed up by developments such at Livingston, or the proposed Drumshoreland new housing. It has come of age, it has proved its resilience and independence and enters the new Millennium with its future assured.

Pumpherston Population Figures

Year	Population
1841	33
1851	50
1861	53
1871	109
1881	93
1885	500

The above figures include both the village and the farm cottages on the Pumpherston estate.

1891	1,359 (village), 210 households
1901	1,462 (village), 1,536 (parish ward)
1911	1,605 (parish ward)
1921	1,529 (parish ward)
1931	1,455 (County Council Electoral Division)
1941	No Census because of the War
1951	1,193 (village) (1,423 County Council Electoral Division)
1961	1,473 (County Council Electoral Division)
1971	1,655 (County Council Electoral Division)
1981	1,321 (village)
1997	1,341 (village), 561 households
2000	1,301 (village)

Further Reading

Vic Armstrong – History of Pumpherston (unpublished)
Stuart Borrowman – Capital of Silicon Glen: West Lothian: transformed for good? (2000)
Alexander Campbell – Cauther Fair: poems and songs (1947)
(Alexander Campbell (1870-1941) was employed for 45 years at Pumpherston Oil Works.)
Andrew Duncan – Historical Notices of the United Presbyterian Congregation of Midcalder (1874)
Alistair Findlay – Shale Voices (1999)
Patrick Gallagher – My Story – by Paddy the Cope (1979)
David Kerr – Shale Oil: Scotland (2nd revised edition 1999)
Hardy Bertram McCall – The History and Antiquities of the Parish of Mid Calder (1894)
John H. McKay – A Social History of the Scottish Shale Oil Industry, 1850-1914 (1984)
Ordnance Survey Namebooks, Parish of Mid Calder (1850s)
John Sommers – History of the Parish of Mid Calder (1838)
Vicky Whyte – Riches below the Castle: a short history of the Pumpherston Oil Company 1883-1993 (1993?)
Files of the *West Lothian Courier*, 1873 to date
Minutes of the Parochial Board and Parish Council of Mid Calder

All of these are available at the West Lothian Local History Library, Library HQ, Hopefield Road, Blackburn, West Lothian.

Index

PLACES

Some other books published by **LUATH** PRESS

SOCIAL HISTORY

Shale Voices

Alistair Findlay
foreword by Tam Dalyell MP
ISBN 0 946487 63 4 PBK £10.99
ISBN 0 946487 78 2 HBK £17.99

'He was at Addiewell oil works. Anyone goes in there is there for keeps.' JOE, Electrician

'There's shale from here to Ayr, you see.' DICK, a Drawer

'The way I describe it is, you're a coal miner and I'm a shale miner. You're a tramp and I'm a toff.' HARRY, a Drawer

'There were sixteen or eighteen Simpsons... ...She was having one every dividend we would say.' SISTERS, from Broxburn

Shale Voices offers a fascinating insight into shale mining, an industry that employed generations of Scots, had an impact on the social, political and cultural history of Scotland and gave birth to today's large oil companies. Author Alistair Findlay was born in the shale mining village of Winchburgh and is the fourth son of a shale miner, Bob Findlay, who became editor of the *West Lothian Courier*. *Shale Voices* combines oral history, local journalism and family history. The generations of communities involved in shale mining provide, in their own words, a unique documentation of the industry and its cultural and political impact.

Photographs, drawings, poetry and short stories make this a thought provoking and entertaining account. It is as much a joy to dip into and feast the eyes on as to read from cover to cover.

'Alistair Findlay has added a basic source material to the study of Scottish history that is invaluable and will be of great benefit to future generations. Scotland owes him a debt of gratitude for undertaking this work.'
TAM DALYELL MP

'...lovingly evoked ...isn't an idle intellectual exercise ...laid out in poetic form, captures the music of speech ...love & respect shines through in this book ...one of the finest pieces of social history I've ever read.'
MARK STEVEN, THE SCOTTISH CONNECTION, BBC RADIO SCOTLAND

'...for thousands of people across the country their attitudes, lifestyles and opinions have been formed through an industry which was once the envy of the world ...captures the essence of the feelings of the time.'
LINDSAY GOULD,
WEST LOTHIAN COURIER

'...the mighty shale bings of West Lothian seem to be a secret which remarkably few outsiders share. how beautifully their russet grit glows in dawn or evening light.'
ANGUS CALDER, THE SCOTSMAN

'Findlay records their voices, as sharp and red as the rock they worked ...their voices are also, in a strange way, freed. Findlay, himself a poet, lays them out on the page as poetry to capture the "dynamics of conversation". The result is to recreate the directness, simplicity and power of everyday speech.'
JOHN FOSTER, THE MORNING STAR

'...the real and rounded history of the people ...important, informative, captivating and inspiring, speckled with hardship and humour, it is well worth a read.'
JOHN STEVENSON,
SCOTLAND IN UNISON

'...the class solidarity and sense of sharing with neighbours in good times and bad could enhance the world of today. Alistair Findlay says it much better than I can... do you not feel echoes of Lewis Grassic Gibbon's Sunset Song in this man's writing?'
WILLIAM WOLFE,
SCOTS INDEPENDENT

A Word for Scotland

Jack Campbell

with a foreword by Magnus Magnusson

ISBN 0 946487 48 0 PBK £12.99

'A word for Scotland' was Lord Beaverbrook's hope when he founded the *Scottish Daily Express*. That word for Scotland quickly became, and was for many years, the national newspaper of Scotland.

The pages of *A Word For Scotland* exude warmth and a wry sense of humour. Jack Campbell takes us behind the scenes to meet the larger-than-life characters and ordinary people who made and recorded the stories. Here we hear the stories behind the stories that hit the headlines in this great yarn of journalism in action.

It would be true to say 'all life is here'. From the Cheapside Street fire of which cost the lives of 19 Glasgow firemen, to the theft of the Stone of Destiny, to the lurid exploits of serial killer Peter Manuel, to encounters with world boxing champions Benny Lynch and Cassius Clay - this book offers telling glimpses of the characters, events, joy and tragedy which make up Scotland's story in the 20th century.

'As a rookie reporter you were proud to work on it and proud to be part of it - it was fine newspaper right at the heartbeat of Scotland.'

RONALD NEIL, Chief Executive of BBC Production, and a reporter on the *Scottish Daily Express* (1963-68)

'This book is a fascinating reminder of Scottish journalism in its heyday. It will be read avidly by those journalists who take pride in their profession – and should be compulsory reading for those who don't.'

JACK WEBSTER, columnist on *The Herald* and *Scottish Daily Express* journalist (1960-80)

The Crofting Years

Francis Thompson

ISBN 0 946487 06 5 PBK £6.95

Crofting is much more than a way of life. It is a storehouse of cultural, linguistic and moral values which holds together a scattered and struggling rural population. This book fills a blank in the written history of crofting over the last two centuries. Bloody conflicts and gunboat diplomacy, treachery, compassion, music and story: all figure in this mine of information on crofting in the Highlands and Islands of Scotland.

'I would recommend this book to all who are interested in the past, but even more so to those who are interested in the future survival of our way of life and culture'
STORNOWAY GAZETTE

'The book is a mine of information on many aspects of the past, among them the homes, the food, the music and the medicine of our crofting forebears.'
John M Macmillan, erstwhile CROFTERS COMMISSIONER FOR LEWIS AND HARRIS

HISTORY

Reportage Scotland: History in the Making

Louise Yeoman

Foreword by Professor David Stevenson

ISBN 0 946487 61 8 PBK £9.99

Events – both major and minor – as seen and recorded by Scots throughout history.

Which king was murdered in a sewer?

What was Dr Fian's love magic?
Who was the half-roasted abbot?
Which cardinal was salted and put in a barrel?
Why did Lord Kitchener's niece try to blow up Burns's cottage?

The answers can all be found in this eclectic mix covering nearly 2000 years of Scottish history. Historian Louise Yeoman's rummage through the manuscript, book and newspaper archives of the National Library of Scotland has yielded an astonishing range of material from a letter to the king of the Picts to in Mary Queen of Scots' own account of the murder of David Riccio; from the execution of William Wallace to accounts of anti-poll tax actions and the opening of the new Scottish Parliament. The book takes pieces from the original French, Latin, Gaelic and Scots and makes them accessible to the general reader, often for the first time.

The result is compelling reading for anyone interested in the history that has made Scotland what it is today.

'Marvellously illuminating and wonderfully readable'. Angus Calder, SCOTLAND ON SUNDAY

'A monumental achievement in drawing together such a rich historical harvest' Chris Holme, THE HERALD

Blind Harry's Wallace

William Hamilton of Gilbertfield

Introduced by Elspeth King

ISBN 0 946487 43 X HBK £15.00
ISBN 0 946487 33 2 PBK £8.99

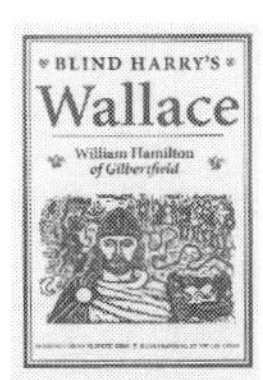

The original story of the real braveheart, Sir William Wallace. Racy, blood on every page, violently anglophobic, grossly embellished, vulgar and disgusting, clumsy and stilted, a literary failure, a great epic.

Whatever the verdict on BLIND HARRY, this is the book which has done more than any other to frame the notion of Scotland's national identity. Despite its numerous 'historical inaccuracies', it remains the principal source for what we now know about the life of Wallace.

The novel and film *Braveheart* were based on the 1722 Hamilton edition of this epic poem. Burns, Wordsworth, Byron and others were greatly influenced by this version 'wherein the old obsolete words are rendered more intelligible', which is said to be the book, next to the Bible, most commonly found in Scottish households in the eighteenth century. Burns even admits to having 'borrowed... a couplet worthy of Homer' directly from Hamilton's version of BLIND HARRY to include in *'Scots wha hae'*.

Elspeth King, in her introduction to this, the first accessible edition of BLIND HARRY in verse form since 1859, draws parallels between the situation in Scotland at the time of Wallace and that in Bosnia and Chechnya in the 1990s. Seven hundred years to the day after the Battle of Stirling Bridge, the 'Settled Will of the Scottish People' was expressed in the devolution referendum of 11 September 1997. She describes this as a landmark opportunity for mature reflection on how the nation has been shaped, and sees BLIND HARRY'S WALLACE as an essential and compelling text for this purpose.

'A true bard of the people'.

TOM SCOTT, THE PENGUIN BOOK OF SCOTTISH VERSE, on Blind Harry.

'A more inventive writer than Shakespeare'.

RANDALL WALLACE

'The story of Wallace poured a Scottish prejudice in my veins which will boil along until the floodgates of life shut in eternal rest'. ROBERT BURNS

'Hamilton's couplets are not the best poetry you will ever read, but they rattle along at a fair pace. In re-issuing this work, the publishers have re-opened the spring from which most of our conceptions of the Wallace legend come'.
SCOTLAND ON SUNDAY

'The return of Blind Harry's Wallace, a man who makes Mel look like a wimp'. THE SCOTSMAN

Old Scotland New Scotland

Jeff Fallow

ISBN 0 946487 40 5 PBK £6.99

'Together we can build a new Scotland based on Labour's values.' DONALD DEWAR, Party Political Broadcast

'Despite the efforts of decent Mr Dewar, the voters may yet conclude they are looking at the same old hacks in brand new suits.'
IAN BELL, *The Independent*

'At times like this you suddenly realise how dangerous the neglect of Scottish history in our schools and universities may turn out to be.' MICHAEL FRY, *The Herald*

'...one of the things I hope will go is our chip on the shoulder about the English... The SNP has a huge responsibility to articulate Scottish independence in a way that is pro-Scottish and not anti-English.'
ALEX SALMOND, *The Scotsman*

Scottish politics have never been more exciting. In *old Scotland new Scotland* Jeff Fallow takes us on a graphic voyage through Scotland's turbulent history, from earliest times through to the present day and beyond. This fast-track guide is the quick way to learn what your history teacher didn't tell you, essential reading for all who seek an understanding of Scotland and its history.

Eschewing the romanticisation of his country's past, Fallow offers a new perspective on an old nation. *'Too many people associate Scottish history with tartan trivia or outworn romantic myth. This book aims to blast that stubborn idea.'*
JEFF FALLOW

Notes from the North
incorporating a Brief History of the Scots and the English

Emma Wood

ISBN 0 946487 46 4 PBK £8.99

Notes on being English
Notes on being in Scotland
Learning from a shared past

Sickened by the English jingoism that surfaced in rampant form during the 1982 Falklands War, Emma Wood started to dream of moving from her home in East Anglia to the Highlands of Scotland. She felt increasingly frustrated and marginalised as Thatcherism got a grip on the southern English psyche. The Scots she met on frequent holidays in the Highlands had no truck with Thatcherism, and she felt at home with grass-roots Scottish anti-authoritarianism. The decision was made. She uprooted and headed for a new life in the north of Scotland.

'An intelligent and perceptive book... calm, reflective, witty and sensitive. It should certainly be read by all English visitors to Scotland, be they tourists or incomers. And it should certainly be read by all Scots concerned about what kind of nation we live in. They might learn something about themselves.' THE HERALD

'... her enlightenment is evident on every page of this perceptive, provocative book.'
MAIL ON SUNDAY

Edinburgh's Historic Mile

Duncan Priddle

ISBN 0 946487 97 9 PBK £2.99

This ancient thoroughfare runs downwards and eastwards for just over a mile. Its narrow closes and wynds, each with a distinctive atmosphere and character, have their own stories to tell. From the looming fortress of the Castle at the top, to the Renaissance beauty of the palace at the bottom, every step along this ancient highway brings the city's past to life – a past both glorious and gory.

Written with all the knowledge and experience the Witchery Tours have gathered in 15 years, it is full of quirky, fun and fascinating stories that you wont find anywhere else.

Designed to fit easily in pocket or bag and with a comprehensive map on the back cover this is the perfect book to take on a walk in Edinburgh or read before you arrive.

Let's Explore Edinburgh Old Town

Anne Bruce English
Illustrations by Cinders McLeod

ISBN 0 946487 98 7 PBK £4.99

The Old Town of Edinburgh has everything. At the highest point is a huge castle. At the foot of the hill there's a palace.

Between them are secret gardens, a museum full of toys, a statue of the world famous Greyfriars Bobby, and much more besides.

There were murders here too (think of Burke and Hare). There's mystery - is preacher John Knox really buried under parking space 44? And then there are the ghosts of Mary King's Close.

You can find out about all this and more in this guide. Read the tales of the Old Town, check out the short quizzes and the Twenty Questions page (all the answers are given), and you'll have plenty to see and do. Join Anne and Cinders on a fascinating and fun journey through time.

The Quest for Arthur

Stuart McHardy

ISBN 1 84282 012 5 HBK £16.99

King Arthur of Camelot and the Knights of the Round Table are enduring romantic figures. A national hero for the Bretons, the Welsh and the English alike Arthur is a potent figure for many. This quest leads to a radical new interpretation of the ancient myth.

Historian, storyteller and folklorist Stuart McHardy believes he has uncovered the origins of this inspirational figure, the true Arthur. He incorporates knowledge of folklore and placename studies with an archaeological understanding of the 6th century.

Combining knowledge of the earliest records and histories of Arthur with an awareness of the importance of oral traditions, this quest leads to the discovery that the enigmatic origins of Arthur lie not in Brittany or England or Wales. Instead they lie in that magic land the ancient Welsh called Y Gogledd, the North; the North of Britain which we now call Scotland.

ON THE TRAIL OF

On the Trail of William Wallace

David R. Ross

ISBN 0 946487 47 2 PBK £7.99

How close to reality was *Braveheart*?

Where was Wallace actually born?

What was the relationship between Wallace and Bruce?

Are there any surviving eye-witness accounts of Wallace?

How does Wallace influence the psyche of today's Scots?

On the Trail of William Wallace offers a refreshing insight into the life and heritage of the great Scots hero whose proud story is at the very heart of what it means to be Scottish. Not concentrating simply on the hard historical facts of Wallace's life, the book also takes into account the real significance of Wallace and his effect on the ordinary Scot through the ages, manifested in the many sites where his memory is marked.

In trying to piece together the jigsaw of the reality of Wallace's life, David Ross weaves a subtle flow of new information with his own observations. His engaging, thoughtful and at times amusing narrative reads with the ease of a historical novel, complete with all the intrigue, treachery and romance required to hold the attention of the casual reader and still entice the more knowledgable historian.

74 places to visit in Scotland and the north of England

One general map and 3 location maps

Stirling and Falkirk battle plans

Wallace's route through London

Chapter on Wallace connections in North America and elsewhere

Reproductions of rarely seen illustrations

On the Trail of William Wallace will be enjoyed by anyone with an interest in Scotland, from the passing tourist to the most fervent nationalist. It is an encyclopaedia-cum-guide book, literally stuffed with fascinating titbits not usually on offer in the conventional history book.

David Ross is organiser of and historical adviser to the Society of William Wallace.

'Historians seem to think all there is to be known about Wallace has already been uncovered. Mr Ross has proved that Wallace studies are in fact in their infancy.' ELSPETH KING, Director the the Stirling Smith Art Museum & Gallery, who annotated and introduced the recent Luath edition of *Blind Harry's Wallace.*

'Better the pen than the sword!'

RANDALL WALLACE, author of *Braveheart,* when asked by David Ross how it felt to be partly responsible for the freedom of a nation following the Devolution Referendum.

On the Trail of Robert the Bruce

David R. Ross

ISBN 0 946487 52 9 PBK £7.99

On the Trail of Robert the Bruce charts the story of Scotland's hero-king from his boyhood, through his days of indecision as Scotland suffered under the English yoke, to his assumption of the crown exactly six months after the death of William Wallace. Here is the astonishing blow by blow account of how, against fearful odds, Bruce led the Scots to win their greatest ever victory. Bannockburn was not the end of the story. The war against English oppression lasted another fourteen years. Bruce lived just long enough to see his dreams of an independent Scotland come to fruition in 1328 with the signing of the Treaty of Edinburgh. The trail takes us to Bruce sites in Scotland, many of the little known and forgotten battle sites in northern England, and as far afield as the Bruce monuments in Andalusia and Jerusalem.

67 places to visit in Scotland and elsewhere.

One general map, 3 location maps and a map of Bruce-connected sites in Ireland.

Bannockburn battle plan.

Drawings and reproductions of rarely seen illustrations.

On the Trail of Robert the Bruce is not all blood and gore. It brings out the love and laughter, pain and passion of one of the great eras of Scottish history. Read it and you will understand why David Ross has never knowingly killed a spider in his life. Once again, he proves himself a master of the popular brand of hands-on history that made *On the Trail of William Wallace* so popular.

'David R. Ross is a proud patriot and unashamed romantic.'

SCOTLAND ON SUNDAY

'Robert the Bruce knew Scotland, knew every class of her people, as no man who ruled her before or since has done. It was he who asked of her a miracle - and she accomplished it.'

AGNES MUIR MACKENZIE

On the Trail of Mary Queen of Scots

J. Keith Cheetham

ISBN 0 946487 50 2 PBK £7.99

Life dealt Mary Queen of Scots love, intrigue, betrayal and tragedy in generous measure.

On the Trail of Mary Queen of Scots traces the major events in the turbulent life of the beautiful, enigmatic queen whose romantic reign and tragic destiny exerts an undimmed fascination over 400 years after her execution.

Places of interest to visit – 99 in Scotland, 35 in England and 29 in France.

One general map and 6 location maps.

Line drawings and illustrations.

Simplified family tree of the royal houses of Tudor and Stuart.

Key sites include:

Linlithgow Palace – Mary's birthplace, now a magnificent ruin

Stirling Castle – where, only nine months old, Mary was crowned Queen of Scotland

Notre Dame Cathedral – where, aged fifteen, she married the future king of France

The Palace of Holyroodhouse – Rizzio, one of Mary's closest advisers, was murdered here and some say his blood still stains the spot where he was stabbed to death

Sheffield Castle – where for fourteen years she languished as prisoner of her cousin, Queen Elizabeth I

Fotheringhay – here Mary finally met her death on the executioner's block.

On the Trail of Mary Queen of Scots is for everyone interested in the life of perhaps the most romantic figure in Scotland's history; a thorough guide to places connected with Mary, it is also a guide to the complexities of her personal and public life.

'In my end is my beginning'
MARY QUEEN OF SCOTS

'...the woman behaves like the Whore of Babylon' JOHN KNOX

On the Trail of Robert Service

GW Lockhart

ISBN 0 946487 24 3 PBK £7.99

Robert Service is famed world-wide for his eye-witness verse-pictures of the Klondike goldrush. As a war poet, his work outsold Owen and Sassoon, and he went on to become the world's first million selling poet. In search of adventure and new experiences, he emigrated from Scotland to Canada in 1890 where he was caught up in the aftermath of the raging gold fever. His vivid dramatic verse bring to life the wild, larger than life characters of the gold rush Yukon, their bar-room brawls, their lust for gold, their trigger-happy gambles with life and love. 'The Shooting of Dan McGrew' is perhaps his most famous poem:

> *A bunch of the boys were whooping it up in the Malamute saloon;*
> *The kid that handles the music box was hitting a ragtime tune;*
> *Back of the bar in a solo game, sat Dangerous Dan McGrew,*
> *And watching his luck was his light o'love, the lady that's known as Lou.*

His storytelling powers have brought Robert Service enduring fame, particularly in North America and Scotland where he is something of a cult figure.

Starting in Scotland, *On the Trail of Robert Service* follows Service as he wanders through British Columbia, Oregon, California, Mexico, Cuba, Tahiti, Russia, Turkey and the Balkans, finally 'settling' in France.

This revised edition includes an expanded selection of illustrations of scenes from the Klondike as well as several photographs from the family of Robert Service on his travels around the world.

Wallace Lockhart, an expert on Scottish traditional folk music and dance, is the author of *Highland Balls & Village Halls* and *Fiddles & Folk*. His relish for a well-told tale in popular vernacular led him to fall in love with the verse of Robert Service and write his biography.

'A fitting tribute to a remarkable man - a bank clerk who wanted to become a cowboy. It is hard to imagine a bank clerk writing such lines as:
A bunch of boys were whooping it up...
The income from his writing actually exceeded his bank salary by a factor of five and he resigned to pursue a full time writing career.' Charles Munn,
THE SCOTTISH BANKER

'Robert Service claimed he wrote for those who wouldnit be seen dead reading poetry. His was an almost unbelievably mobile life... Lockhart hangs on breathlessly, enthusiastically unearthing clues to the poet's life.' Ruth Thomas,
SCOTTISH BOOK COLLECTOR

'This enthralling biography will delight Service lovers in both the Old World and the New.' Marilyn Wright,
SCOTS INDEPENDENT

On the Trail of John Muir

Cherry Good

ISBN 0 946487 62 6 PBK £7.99

Follow the man who made the US go green. Confidant of presidents, father of American National Parks, trailblazer of world conservation and voted a Man of the Millennium in the US, John Muir's life and work is of continuing relevance. A man ahead of his time who saw the wilderness he loved threatened by industrialisation and determined to protect it, a crusade in which he was largely successful. His love of the wilderness began at an early age and he was filled with wanderlust all his life.

'Only by going in silence, without baggage, can on truly get into the heart of the wilderness. All other travel is mere dust and hotels and baggage and chatter.' JOHN MUIR

Braving mosquitoes and black bears Cherry Good set herself on his trail – Dunbar, Scotland; Fountain Lake and Hickory Hill, Wisconsin; Yosemite Valley and the Sierra Nevada, California; the Grand Canyon, Arizona; Alaska; and Canada – to tell his story. John Muir was himself a prolific writer, and Good draws on his books, articles, letters and diaries to produce an account that is lively, intimate, humorous and anecdotal, and that provides refreshing new insights into the hero of world conservation.

John Muir chronology
General map plus 10 detailed maps covering the US, Canada and Scotland
Original colour photographs
Afterword advises on how to get involved
Conservation websites and addresses

Muir's importance has long been acknowledged in the US with over 200 sites of scenic beauty named after him. He was a Founder of The Sierra Club which now has over ½ million members. Due to the movement he started some 360 million acres of wilderness are now protected. This is a book which shows Muir not simply as a hero but as likeable humorous and self-effacing man of extraordinary vision.

'I do hope that those who read this book will burn with the same enthusiasm for John Muir which the author shows.'
WEST HIGHLAND FREE PRESS

On the Trail of Robert Burns

John Cairney

ISBN 0 946487 51 0 PBK £7.99

Is there anything new to say about Robert Burns?

John Cairney says it's time to trash Burns the Brand and come on the trail of the real Robert Burns. He is the best of travelling companions on this convivial, entertaining journey to the heart of the Burns story.

Internationally known as 'the face of Robert Burns', John Cairney believes that the traditional Burns tourist trail urgently needs to find a new direction. In an acting career spanning forty years he has often lived and breathed Robert Burns on stage. *On the*

Trail of Robert Burns shows just how well he can get under the skin of a character. This fascinating journey around Scotland is a rediscovery of Scotland's national bard as a flesh and blood genius.

On the Trail of Robert Burns outlines five tours, mainly in Scotland. Key sites include:

Alloway - Burns' birthplace. 'Tam O' Shanter' draws on the witch-stories about Alloway Kirk first heard by Burns in his childhood.
Mossgiel - between 1784 and 1786 in a phenomenal burst of creativity Burns wrote some of his most memorable poems including 'Holy Willie's Prayer' and 'To a Mouse.'
Kilmarnock - the famous Kilmarnock edition of *Poems Chiefly in the Scottish Dialect* published in 1786.
Edinburgh - fame and Clarinda (among others) embraced him.
Dumfries - Burns died at the age of 37. The trail ends at the Burns mausoleum in St Michael's churchyard.

'For me an aim I never fash
I rhyme for fun'.
ROBERT BURNS

'My love affair on stage with Burns started in London in 1959. It was consumated on stage at the Traverse Theatre in Edinburgh in 1965 and has continued happily ever since'.

JOHN CAIRNEY

'The trail is expertly, touchingly and amusingly followed'.
THE HERALD

On the Trail of Bonnie Prince Charlie

David R. Ross

ISBN 0 946487 68 5 PBK £7.99

On the Trail of Bonnie Prince Charlie is the story of the Young Pretender. Born in Italy, grandson of James VII, at a time when the German house of Hanover was on the throne, his father was regarded by many as the righful king. Bonnie Prince Charlie's campaign to retake the throne in his father's name changed the fate of Scotland. The Jacobite movement was responsible for the '45 Uprising, one of the most decisive times in Scottish history. The suffering following the battle of Culloden in 1746 still evokes emotion. Charles' own journey immediately after Culloden is well known: hiding in the heather, escaping to Skye with Flora MacDonald. Little known of is his return to London in 1750 incognito, where he converted to Protestantism (he re-converted to Catholicism before he died and is buried in the Vatican). He was often unwelcome in Europe after the failure of the uprising and came to hate any mention of Scotland and his lost chance.

79 places to visit in Scotland and England
One general map and 4 location maps
Prestonpans, Clifton, Falkirk and Culloden battle plans
Simplified family tree
Rarely seen illustrations

Yet again popular historian David R. Ross brings his own style to one of Scotland's most famous figures. Bonnie Prince Charlie is part of the folklore of Scotland. He brings forth feelings of antagonism from some and romanticism from others, but all agree on his legal right to the throne.

Knowing the story behind the place can bring the landscape to life. Take this book with you on your travels and follow the route taken by Charles' forces on their doomed march.

'Ross writes with an immediacy, a dynamism, that makes his subjects come alive on the page.'

DUNDEE COURIER

On the Trail of Queen Victoria in the Highlands

Ian R. Mitchell

UK ISBN 0 946487 79 0 PBK £7.99

How many Munros did Queen Victoria bag?

What 'essential services' did John Brown perform for Victoria?

(and why was Albert always tired?)

How many horses (to the nearest hundred) were needed to undertake a Royal Tour?

What happens when you send a republican on the tracks of Queen Victoria in the Highlands?

a.. you get a book somewhat more interesting than the usual run of the mill royalist biographies!

Ian R. Mitchell took up the challenge of attempting to write with critical empathy on the peregrinations of Vikki Regina in the Highlands, and about her residence at Balmoral, through which a neo-feudal fairyland was created on Upper Deeside. The expeditions, social rituals and iconography of that world are explored and exploded from within, in what Mitchell terms a Bolshevisation of Balmorality. He follows in Victoria's footsteps throughout the Cairngorms and beyond, to the further reaches of the Highlands. On this journey, a grudging respect and even affection for Vikki ('the best of the bunch') emerges.

The book is designed to enable the armchair/motorised reader, or walker, to follow in the steps of the most widely-travelled royal personage in the Highlands since Bonnie Prince Charlie had wandered there a century earlier.

Index map and 12 detailed maps

21 walks in Victoria's footsteps

Rarely seen Washington Wilson photographs

Colour and black and white reproductions of contemporary paintings

On the Trail of Queen Victoria in the Highlands will also appeal to those with an interest in the social and cultural history of Scotland and the Highlands - and the author, ever-mindful of his own 'royalties', hopes the declining band of monarchists might also be persuaded to give the book a try.

SPORT

Pilgrims In The Rough: St Andrews beyond the 19th hole

Michael Tobert

ISBN 0 946487 74 X PBK £7.99

'A travel book about St. Andrews. A book that combines the game I love and the course I have played for 20 years, with the town that I consider as close to paradise as I am likely to find on this side of the pearly gates.' MICHAEL TOBERT

With ghosts, witches and squabbling clerics, *Pilgrims in the Rough* is a funny and affectionate portrayal of Michael Tobert's home town. The author has always wanted to write a travel book – but he has done more than that. Combining tourist information with history, humour and anecdote, he has written a book that will appeal to golfer and non golfer, local and visitor, alike.

While *Pilgrims in the Rough* is more than just a guide to clubs and caddies, it is nonetheless packed with information for the golf enthusiast. It features a detailed map of the course and the low down from a regular St Andrews player on booking times, the clubs and each of the holes on the notorious Old Course.

The book also contains an informative guide to the attractions of the town and the best places to stay and to eat out. Michael Tobert's infectious enthusiasm for St Andrews will even persuade the most jaded golf widow or widower that the town is worth a visit!

'An extraordinary book' THE OBSERVER

'Tobert displays genuine erudition on such topics as the history of the cathedral and university and, of course, the tricky business of playing the Old Course itself.' THE SCOTSMAN

Over the Top with the Tartan Army (Active Service 1992-97)

Andrew McArthur

ISBN 0 946487 45 6 PBK £7.99

Scotland has witnessed the growth of a new and curious military phenomenon – grown men bedecked in tartan yomping across the globe, hell-bent on benevolence and ritualistic bevvying. What noble cause does this famous army serve? Why, football of course!

Taking us on an erratic world tour, McArthur gives a frighteningly funny insider's eye view of active service with the Tartan Army – the madcap antics of Scotland's travelling support in the '90s, written from the inside, covering campaigns and skirmishes from Euro '92 up to the qualifying drama for France '98 in places as diverse as Russia, the Faroes, Belarus, Sweden, Monte Carlo, Estonia, Latvia, USA and Finland.

This book is a must for any football fan who likes a good laugh.

'I commend this book to all football supporters'. Graham Spiers, SCOTLAND ON SUNDAY

'In wishing Andy McArthur all the best with this publication, I do hope he will be in a position to produce a sequel after our participation in the World Cup in France'. CRAIG BROWN, Scotland Team Coach

All royalties on sales of the book are going to Scottish charities.

FICTION

The Bannockburn Years

William Scott

ISBN 0 946487 34 0 PBK £7.95

A present day Edinburgh solicitor stumbles across reference to a document of value to the Nation State of Scotland. He tracks down the document on the Isle of Bute, a document which probes the real 'quaestiones' about nationhood and national identity. The document ends up being published, but is it authentic and does it matter? Almost 700 years on, these 'quaestiones' are still worth asking.

Written with pace and passion, William Scott has devised an intriguing vehicle to open up new ways of looking at the future of Scotland and its people. He presents an alternative interpretation of how the Battle of Bannockburn was fought, and through the Bannatyne manuscript he draws the reader into the minds of those involved.

Winner of the 1997 Constable Trophy, the premier award in Scotland for an unpublished novel, this book offers new insights to both the academic and the general reader which are sure to provoke further discussion and debate.

'A brilliant storyteller. I shall expect to see your name writ large hereafter.' NIGEL TRANTER.

'... a compulsive read.' PH Scott, THE SCOTSMAN

The Great Melnikov

Hugh Maclachlan

ISBN 0 946487 42 1 PBK £7.95

A well crafted, gripping novel, written in a style reminiscent of John Buchan and set in London and the Scottish Highlands during the First World War, *The Great Melnikov* is a dark tale of double-cross and deception. We first meet Melnikov, one-time star of the German circus, languishing as a down-and-out in Trafalgar Square. He soon finds himself drawn into a tortuous web of intrigue. He is a complex man whose personal struggle with alcoholism is an inner drama which parallels the tense twists and turns as a spy mystery unfolds. Melnikov's options are narrowing. The circle of threat is closing. Will Melnikov outwit the sinister enemy spy network? Can he summon the will and the wit to survive?

Hugh Maclachlan, in his first full length novel, demonstrates an undoubted ability to tell a good story well. His earlier stories have been broadcast on Radio Scotland, and he has the rare distinction of being shortlisted for the Macallan/Scotland on Sunday Short Story Competition two years in succession.

'... a satisfying rip-roarer of a thriller... an undeniable page turner, racing along to a suitably cinematic ending, richly descriptive yet clear and lean.'

THE SCOTSMAN

But n Ben A-Go-Go

Matthew Fitt

ISBN 0 946487 82 0 HBK £10.99
ISBN 1 84282 014 1 PBK £6.99

The year is 2090. Global flooding has left most of Scotland under water. The descendants of those who survived God's Flood live in a community of floating island parishes, known collectively as Port.

Port's citizens live in mortal fear of Senga, a supervirus whose victims are kept in a giant hospital warehouse in sealed capsules called Kists.

Paolo Broon is a low-ranking cyberjanny. His life-partner, Nadia, lies forgotten and alone in Omega Kist 624 in the Rigo Imbeki Medical Center. When he receives an unexpected message from his radge criminal father to meet him at But n Ben A-Go-Go, Paolo's life is changed forever.

He must traverse VINE, Port and the Drylands and deal with rebel American tourists and crabbit Dundonian microchips to discover the truth about his family's past in order to free Nadia from the sair grip of the merciless Senga.

Set in a distinctly unbonnie future-Scotland, the novel's dangerous atmosphere and psychologically-malkied characters weave a tale that both chills and intrigues. In *But n Ben A-Go-Go* Matthew Fitt takes the allegedly dead language of Scots and energises it with a narrative that crackles and fizzes with life.

'After an initial shock, readers of this sprightly and imaginative tale will begin to relish its verbal impetus, where a standard Lallans, laced with bits of Dundonian and Aberdonian, is

stretched and skelped to meet the demands of cyberjannies and virtual hoorhooses.

Eurobawbees, rooburgers, mutant kelpies, and titanic blooters from supertyphoons make sure that the Scottish peninsula is no more parochial than its language. I recommend an entertaining and ground-breaking book.'

EDWIN MORGAN

'Matthew Fitt's instinctive use of Scots is spellbinding. This is an assured novel of real inventiveness. Be prepared to boldly go...' ELLIE McDONALD

'Easier to read than Shakespeare – wice the fun.' DES DILLON

The Strange Case of R L Stevenson

Richard Woodhead

ISBN 0 946487 86 3 HBK £16.99

A consultant physician for 22 years with a strong interest in Robert Louis Stevenson's life and work, Richard Woodhead was intrigued by the questions raised by the references to his symptoms. The assumption that he suffered from consumption – the diagnosis of the day – is challenged here. Consumption (tuberculosis), a scourge of nineteenth century society, it was regarded as severely debilitating if not a death sentence. Dr Woodhead examines how Stevenson's life was affected by his illness and his perception of it.

This fictional work puts words into the mouths of five doctors who treated RLS at different periods of his adult life. Though these doctors existed in real-life, little is documented of their private conversations with RLS. However everything Dr Woodhead postulates could have occurred within the known framework of RLS's life. Detailed use of Stevenson's own writing adds authenticity to the views espoused in the book.

RLS's writing continues to compel readers today. The fact that he did much of his writing while confined to his sickbed is fascinating. What illness could have contributed to his creativity?

POETRY

Poems to be read aloud

Collected and with an introduction by Tom Atkinson

ISBN 0 946487 00 6 PBK £5.00

This personal collection of doggerel and verse ranging from the tear-jerking *Green Eye of the Yellow God* to the rarely printed, bawdy *Eskimo Nell* has a lively cult following. Much borrowed and rarely returned, this is a book for reading aloud in very good company, preferably after a dram or twa. You are guaranteed a warm welcome if you arrive at a gathering with this little volume in your pocket.

Scots Poems to be Read Aloud

Collectit an wi an innin by

Stuart McHardy

ISBN 0 946487 81 2 PBK £5.00

This personal collection of well-known and not-so-well-known Scots poems to read aloud includes great works of art and simple pieces of questionable 'literary merit'.

With an emphasis on humour it's a great companion volume to Tom Atkinson's *Poems to be Read Aloud*. For those who love poetry its a wonderful anthology to have to hand, and for all those people who do not normally read poetry this collection is for you. It's a book to encourage the tradi-

tional Scottish ceilidh of song and recitation, whether in the home, the local hall or the mountain bothy.

'One of the great strengths of Scots has always been its capacity for strong rhythm and rhyme as well as its sophisticated ability to handle human emotions in all their variety.'

STUART McHARDY

'Scots Poems to be Read Aloud is pure entertainment - at home, on a stag or a hen night, Hogmanay, Burns Night, in fact any party night.' SUNDAY POST

The Luath Burns Companion

John Cairney

ISBN 1 84282 000 1 PBK £10.00

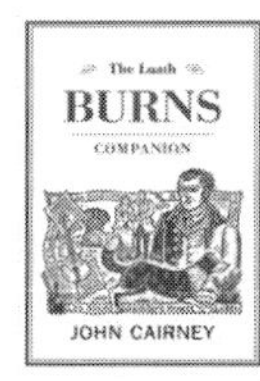

'Robert Burns was born in a thunderstorm and lived his brief life by flashes of lightning.'

So says John Cairney in his introduction. In those flashes his genius revealed itself.

This collection is not another 'complete works' but a personal selection from 'The Man Who Played Robert Burns'. This is very much John's book. His favourites are reproduced here and he talks about them with an obvious love of the man and his work. His depth of knowledge and understanding has been garnered over forty years of study, writing and performance.

The collection includes sixty poems, songs and other works; and an essay that explores Burns's life and influences, his triumphs and tragedies. This informed introduction provides the reader with an insight into Burns's world. Burns's work has drama, passion, pathos and humour. His careful workmanship is concealed by the spontaneity of his verse. He was always a forward thinking man and remains a writer for the future.

Men & Beasts: Wild Men and Tame Animals of Scotland

Poems and Prose by Valerie Gillies

Photographs by Rebecca Marr

ISBN 0 946487 92 8 PBK £15.00

Come and meet some wild men and tame beasts. Explore the fleeting moment and capture the passing of time in these portrait studies which document a year's journey. Travel across Scotland with poet Valerie Gillies and photographer Rebecca Marr: share their passion for a land where wild men can sometimes be tamed and tame beasts can get really wild.

Among the wild men they find are a gunner in Edinburgh Castle, a Highland shepherd, a ferryman on the River Almond, an eel fisher on Loch Ness, a Borders fencer, and a beekeeper on a Lowland estate.

The beasts portrayed in their own settings include Clydesdale foals, Scottish deerhounds, Highland cattle, blackface sheep, falcons, lurchers, bees, pigs, cashmere goats, hens, cockerels, tame swans and transgenic lambs.

Photograph, poem and reportage – a unique take on Scotland today.

'Goin aroon the Borders wi Valerie an' Rebecca did my reputation the world o good. It's no often they see us wi beautiful talented women, ye ken.' WALTER ELLIOT, fencer and historian

'These poems are rooted in the elemental world.' ROBERT NYE, reviewing The Chanter's Tune in The Times

'Valerie Gillies is one of the most original voices of the fertile avant-guarde Scottish poetry.' MARCO FAZZINI, l'Arco, Italia

'The work of Valerie Gillies and Rebecca Marr is the result of true collaboration based on insight, empathy and generosity.' JULIE LAWSON, Studies in Photography

'Rebecca Marr's photos never fall into the trap of mere illustration, but rather they show a very individual vision – creative interpretation rather than prosaic document.' ROBIN GILLANDERS, photographer

Half the royalties genereated from the sale of this publication will go to Maggie's Centre for the care of cancer patients.

'Nothing but Heather!'

Gerry Cambridge

ISBN 0 946487 49 9 PBK £15.00

Enter the world of Scottish nature – bizarre, brutal, often beautiful, always fascinating – as seen through the lens and poems of Gerry Cambridge, one of Scotland's most distinctive contemporary poets.

On film and in words, Cambridge brings unusual focus to bear on lives as diverse as those of dragonflies, hermit crabs, short-eared owls, and wood anemones. The result is both an instructive look by a naturalist at some of the flora and fauna of Scotland and a poet's aesthetic journey.

This exceptional collection comprises 48 poems matched with 48 captioned photographs. In his introduction Cambridge explores the origins of the project and the approaches to nature taken by other poets, and incorporates a wry account of an unwillingly-sectarian, farm-labouring, bird-obsessed adolescence in rural Ayrshire in the 1970s.

'Keats felt that the beauty of a rainbow was somehow tarnished by knowledge of its properties. Yet the natural world is surely made more, not less, marvellous by awareness of its workings. In the poems that accompany these pictures, I have tried to give an inkling of that. May the marriage of verse and image enlarge the reader's appreciation and, perhaps, insight into the chomping, scurrying, quivering, procreating and dying kingdom, however many miles it be beyond the door.'
GERRY CAMBRIDGE

'a real poet, with a sense of the music of language and the poetry of life...' KATHLEEN RAINE

'one of the most promising and original of modern Scottish poets... a master of form and subtlety.'
GEORGE MACKAY BROWN

MUSIC AND DANCE

Highland Balls and Village Halls

GW Lockhart

ISBN 0 946487 12 X PBK £6.95

Acknowledged as a classic in Scottish dancing circles throughout the world. Anecdotes, Scottish history, dress and dance steps are all included in this *'delightful little book, full of interest... both a personal account and an understanding look at the making of traditions.'* NEW ZEALAND SCOTTISH COUNTRY DANCES MAGAZINE

'A delightful survey of Scottish dancing and custom. Informative, concise and opinionated, it guides the reader across the history and geography of country dance and ends by detailing the 12 dances every Scot should know – the most famous being the Eightsome Reel, 'the greatest longest, rowdiest, most diabolically executed of all the Scottish country dances'.' THE HERALD

'A pot-pourri of every facet of Scottish country dancing. It will bring back memories of petronella turns and poussettes and make you eager to take part in a Broun's reel or a dashing white sergeant!' DUNDEE COURIER AND ADVERTISER

'An excellent an very readable insight into the traditions and customs of Scottish country dancing. The author takes us on a tour from his own early days jigging in the village hall to the characters and traditions that have made our own brand of dance popular throughout the world.' SUNDAY POST

Fiddles & Folk: A celebration of the re-emergence of Scotland's musical heritage

GW Lockhart

ISBN 0 946487 38 3 PBK £7.95

In *Fiddles & Folk*, his companion volume to *Highland Balls and Village Halls*, Wallace Lockhart meets up with many of the people who have created the renaissance of Scot-land's music at home and overseas.

From Dougie MacLean, Hamish Henderson, the Battlefield Band, the Whistlebinkies, the Scottish Fiddle Orchestra, the McCalmans and many more come the stories that break down the musical barriers between Scotland's past and present, and between the diverse musical forms which have woven together to create the dynamism of the music today.

'I have tried to avoid a formal approach to Scottish music as it affects those of us with our musical heritage coursing through our veins. The picture I have sought is one of many brush strokes, looking at how some individuals have come to the fore, examining their music, lives, thoughts, even philosophies...'
WALLACE LOCKHART

' "I never had a narrow, woolly-jumper, fingers stuck in the ear approach to music. We have a musical heritage here that is the envy of the rest of the world. Most countries just can't compete," he [Ian Green, Greentrax] says. And as young Scots tire of Oasis and Blur, they will realise that there is a wealth of young Scottish music on their doorstep just waiting to be discovered.' THE SCOTSMAN

For anyone whose heart lifts at the sound of fiddle or pipes, this book takes you on a delightful journey, full of humour and respect, in the company of some of the performers who have taken Scotland's music around the world and come back enriched.

FOOD AND DRINK

The Whisky Muse
Scotch Whisky in Poem and Song

Collected and introduced by ROBIN LAING

Illustrated by BOB DEWAR

ISBN 0 946487 95 2 PBK £12.99

Whisky – the water of life, perhaps Scotland's best known contribution to the world

Muse – goddess of creative endeavour

The Whisky Muse – the spark of inspiration to many of Scotland's great poets and songwriters

This is a collection of the best poems and songs, both old and new, on the subject of that great Scottish love, whisky. Brought together by Robin Laing, a highly respected Scottish folk-singer and songwriter, and based on his one-man show The Angel's

Share, it combines two of his passions – folk song and whisky. Each poem and song is accompanied by fascinating additional information, and the book is full of sundry other interesting tit-bits on the process of whisky-making.

Various themes emerge from Scottish whisky poems and songs that reflect the pleasures (and medicinal benefits) of imbibing this most beloved of spirits as well as the unfortunate consequences of overindulgence, the centuries of religious disapproval, the temperance movement and the exciseman.

The Scots are a musical nation renowned for the warmth of their hospitality and the tendency to assert the superiority of their whisky over any other in the world. The stories told here are lubricated by warmth and companionship as well as a dram. Slainte.

'uno squisito canautore scozzese' LA REPUBBLICA

'I am fascinated by the effects of maturing whisky in oak casks. That is what gives it such character and distinctiveness and makes it the best drink in the world.' ROBIN LAING

'I first met Robin Laing and Bob Dewar within the hallowed halls of the Scotch Malt Whisky Society in Leith, where Robin and I sit on the Nosing Panel which selects casks of malt whisky for bottling, while Bob executes the famous cartoons which illustrate our findings and embellish the Society's Newsletter. The panel's onerous job is made lighter by Robin's ability not only to sniff out elusive scents, but to describe them wittily and accurately, and in this unique collection of ninety-five songs and poems about Scotch whisky he has exercised precisely the same skill of sniffing out treasures. As a highly accomplished singer-songwriter, he also describes them authoritatively, while Bob's illustrations add wit and humour. This splendid book is necessary reading for anyone interested in whisky and song. It encapsulates Scottish folk culture and the very spirit of Scotland.' Charles MacLean, Editor at Large, WHISKY MAGAZINE

Edinburgh and Leith Pub Guide

Stuart McHardy

ISBN 0 946487 80 4 PBK £4.95

You might be in Edinburgh to explore the closes and wynds of one of Europe's most beautiful cities, to sample the finest Scotch whiskies and to discover a rich Celtic heritage of traditional music and story-telling. Or you might be in Leith to get trashed. Either way, this is the guide for you.

With the able assistance of his long time drinking partner, 'the Man from Fife', Stuart McHardy has dragged his tired old frame around over two hundred pubs – all in the name of research, of course. Alongside drinking numerous pints, he has managed to compile enough historical anecdote and practical information to allow anyone with a sturdy liver to follow in his footsteps.

Although Stuart unashamedly gives top marks to his favourite haunts, he rates most highly those pubs that are original, distinctive and cater to the needs of their clientele. Be it domino league or play-station league, pina colada or a pint of heavy, filled foccacia or mince and tatties, Stuart has found a decent pub that does it.

Over 200 pubs

12 pub trails plus maps

Helpful rating system

Brief guide to Scottish beers and whiskies

'The Man from Fife's wry take on each pub

Discover the answers to such essential questions as:

Which pubs are recommended by whisky wholesalers for sampling?

Where can you find a pub that has links with Bonnie Prince Charlie and Mary Queen of Scots?

Which pub serves kangaroo burgers?

Where can you go for a drop of mead in Edinburgh?

Which pub has a toy crocodile in pride of place behind the bar?

How has Stuart survived all these years?

Long familiar with Edinburgh and Leith's drinking dens, watering holes, shebeens and dens of iniquity, Stuart McHardy has penned a bible for the booze connoisseur. Whether you're here for Hogmanay, a Six Nations weekend, the Festival, just one evening or the rest of your life, this is the companion to slip in your pocket or handbag as you venture out in search of the craic.

CURRENT ISSUES

[Un]comfortably Numb: A Prison Requiem

Maureen Maguire

ISBN 1 84282 001 X PBK £8.99

'People may think I've taken the easy way out but please believe me this is the hardest thing I've ever had to do.'

YVONNE GILMOUR

It was Christmas Eve, the atmosphere in Cornton Vale prison was festive, the girls in high spirits as they were locked up for the night. One of their favourite songs, Pink Floyd's *Comfortably Numb*, played loudly from a nearby cell as Yvonne Gilmour wrote her suicide note. She was the sixth of eight inmates to take their own lives in Cornton Vale prison over a short period of time.

[Un]comfortably Numb follows Yvonne through a difficult childhood, a chaotic adolescence and drug addiction to life and death behind bars. Her story is representative of many women in our prisons today. They are not criminals (only 1% are convicted for violent crimes) and two-thirds are between the ages of sixteen and thirty. Suicide rates among them are rising dramatically. Do these vulnerable young girls really belong in prison?

This is a powerful and moving story told in the words of those involved: Yvonne and her family, fellow prisoners, prison officers, social workers, drug workers. It challenges us with questions which demand answers if more deaths are to be avoided – and offers a tragic indictment of social exclusion in 21st century Britain, told in the real voices of women behind bars.

'Uncomfortably Numb is not a legal textbook or a jurisprudential treatise... it is an investigation into something our sophisticated society can't easily face' AUSTIN LAFFERTY

FOLKLORE

Scotland: Myth, Legend and Folklore

Stuart McHardy

ISBN: 0 946487 69 3 PBK 7.99

Who were the people who built the megaliths?

What great warriors sleep beneath the Hollow Hills?

Were the early Scottish saints just pagans in disguise?

Was King Arthur really Scottish?

When was Nessie first sighted?

This is a book about Scotland drawn from hundreds, if not thousands of years of story-telling. From the oral traditions of the Scots, Gaelic and Norse speakers of the past, it presents a new picture of who the Scottish are and where they come from. The stories that McHardy recounts may be hilarious, tragic, heroic, frightening or just plain bizzare, but they all provide an insight into a unique tradition of myth, legend and folklore that has marked both the language and landscape of Scotland.

The Supernatural Highlands

Francis Thompson

ISBN 0 946487 31 6 PBK £8.99

An authoritative exploration of the otherworld of the Highlander, happenings and beings hitherto thought to be outwith the ordinary forces of nature. A simple introduction to the way of life of rural Highland and Island communities, this new edition weaves a path through second sight, the evil eye, witchcraft, ghosts, fairies and other supernatural beings, offering new sightlines on areas of belief once dismissed as folklore and superstition.

Tall Tales from an Island

Peter Macnab

ISBN 0 946487 07 3 PBK £8.99

Peter Macnab was born and reared on Mull. He heard many of these tales as a lad, and others he has listened to in later years.

There are humorous tales, grim tales, witty tales, tales of witchcraft, tales of love, tales of heroism, tales of treachery, historical tales and tales of yesteryear.

A popular lecturer, broadcaster and writer, Peter Macnab is the author of a number of books and articles about Mull, the island he knows so intimately and loves so much. As he himself puts it in his introduction to this book 'I am of the unswerving opinion that nowhere else in the world will you find a better way of life, nor a finer people with whom to share it.'

'All islands, it seems, have a rich store of characters whose stories represent a kind of sub-culture without which island life would be that much poorer. Macnab has succeeded in giving the retelling of the stories a special Mull flavour, so much so that one can visualise the storytellers sitting on a bench outside the house with a few cronies, puffing on their pipes and listening with nodding approval.'

WEST HIGHLAND FREE PRESS

Tales from the North Coast

Alan Temperley

ISBN 0 946487 18 9 PBK £8.99

Seals and shipwrecks, witches and fairies, curses and clearances, fact and fantasy – the authentic tales in this collection come straight from the heart of a small Highland community. Children and adults alike responsd to their timeless appeal. These *Tales of the North Coast* were collected in the early 1970s by Alan Temperley and young people at Farr Secondary School in Sutherland. All the stories were gathered from the area between the Kyle of Tongue and Strath Halladale, in scattered communities wonderfully rich in lore that had been passed on by word of mouth down the generations. This wide-ranging selection provides a satisying balance between intriguing tales of the supernatural and more everyday occurrences. The book also includes chilling eyewitness accounts of the notorious Strathnaver Clearances when tenants were given a few hours to pack up and get out of their homes, which were then burned to the ground.

Underlying the continuity through the generations, this new edition has a foreward by Jim Johnston, the head teacher at Farr, and includes the vigorous linocut images produced by the young people under the guidance of their art teacher, Elliot Rudie.

Since the original publication of this book, Alan Temperley has gone on to become a highly regarded writer for children.

'The general reader will find this book's spontaneity, its pictures by the children and its fun utterly charming.'
SCOTTISH REVIEW

'An admirable book which should serve as an encouragement to other districts to gather what remains of their heritage of folktales.'
SCOTTISH EDUCATION JOURNAL

Luath Storyteller: Highland Myths & Legends

George W. Macpherson

ISBN 1 84282 003 6 PBK £5.00

The mythical, the legendary, the true... This is the stuff of stories and storytellers, the stuff of an age-old tradition in almost every country in the world, and none more so than Scotland. Celtic heroes, Fairies, Druids, Selkies, Sea horses, Magicians, Giants, Viking invaders; all feature in this collection of traditional Scottish tales, the like of which were told round camp fires centuries ago, and are still told today.

George W. Macpherson has dipped into his phenomenal repertoire of tales to compile this diverse collection of traditional stories, designed to be read aloud. Each has been passed from generation to generation, some are two and a half thousand years old.

From the Celtic legends of Cuchullin and Fionn to the mythical tales of seal-people and magicians these stories have a timeless quality. Often, strands of the stories will interweave and cross over, building a delicate tapestry of Scotland as a mystical, enchanted land. *'The result is vivid and impressive, conveying the tragic dignity of the ancient warrior, or the devoted love of the seal woman and her fisher mate. The personalities and circumstances of people long gone are brought fully to life by the power of the storyteller's words. The ancestors take form before us in the visual imagination.'*
DR DONALD SMITH, THE SCOTTISH STORYTELLING CENTRE

'one of Scotland's best storytellers'
WESTDEUTSCHER RUNDFUNK KOHN

'I loved all your stories, some were sad and some were great. We all hope to see you another day'
SARAH BOCCACCIO, ECOLE PRIMAIRE INTERNATIONALE

'This is your genuine article'
MARK FISHER, THE LIST

'Legend and myth join with humour and gentle wit to create a special magic'
JOY HENDRY, SCOTSMAN

LUATH GUIDES TO SCOTLAND

Mull and Iona: Highways and Byways
Peter Macnab
ISBN 0 946487 58 8 PBK £4.95

South West Scotland
Tom Atkinson
ISBN 0 946487 04 9 PBK £4.95

The West Highlands: The Lonely Lands
Tom Atkinson
ISBN 0 946487 56 1 PBK £4.95

The Northern Highlands: The Empty Lands
Tom Atkinson
ISBN 0 946487 55 3 PBK £4.95

The North West Highlands: Roads to the Isles
Tom Atkinson
ISBN 0 946487 54 5 PBK £4.95

WALK WITH LUATH

Mountain Days & Bothy Nights
Dave Brown and Ian Mitchell
ISBN 0 946487 15 4 PBK £7.50

The Joy of Hillwalking
Ralph Storer
ISBN 0 946487 28 6 PBK £7.50

Scotland's Mountains before the Mountaineers
Ian Mitchell
ISBN 0 946487 39 1 PBK £9.99

LUATH WALKING GUIDES

Walks in the Cairngorms
Ernest Cross
ISBN 0 946487 09 X PBK £4.95

Short Walks in the Cairngorms
Ernest Cross
ISBN 0 946487 23 5 PBK £4.95

NATURAL SCOTLAND

Wild Scotland: The essential guide to finding the best of natural Scotland
James McCarthy
Photography by Laurie Campbell
ISBN 0 946487 37 5 PBK £7.50

Scotland Land and People An Inhabited Solitude
James McCarthy
ISBN 0 946487 57 X PBK £7.99

The Highland Geology Trail
John L Roberts
ISBN 0 946487 36 7 PBK £4.99

Rum: Nature's Island
Magnus Magnusson
ISBN 0 946487 32 4 PBK £7.95

Red Sky at Night
John Barrington
ISBN 0 946487 60 X PBK £8.99

Listen to the Trees
Don MacCaskill
ISBN 0 946487 65 0 PBK £9.99

Wildlife: Otters – On the Swirl of the Tide
Bridget MacCaskill
ISBN 0 946487 67 7 PBK £9.99

Wildlife: Foxes – The Blood is Wild
Bridget MacCaskill
ISBN 0 946487 71 5 PBK £9.99

BIOGRAPHY

Tobermory Teuchter: A first-hand account of life on Mull in the early years of the 20th century
Peter Macnab
ISBN 0 946487 41 3 PBK £7.99

Bare Feet and Tackety Boots
Archie Cameron
ISBN 0 946487 17 0 PBK £7.95

The Last Lighthouse
Sharma Kraustopf
ISBN 0 946487 96 0 PBK £7.99

Come Dungeons Dark
John Taylor Caldwell
ISBN 0 946487 19 7 PBK £6.95

POETRY

Caledonian Cramboclink: verse, broadsheets and inconversation
William Neill
ISBN 0 946487 53 7 PBK £8.99

CURRENT ISSUES

Scotland - Land and Power the agenda for land reform
Andy Wightman
foreword by Lesley Riddoch
ISBN 0 946487 70 7 PBK £5.00

Trident on Trial the case for people's disarmament
Angie Zelter
ISBN 1 84282 004 4 PBK £9.99

Broomie Law
Cinders McLeod
ISBN 0 946487 99 5 PBK £4.00

Luath Press Limited

committed to publishing well written books worth reading

LUATH PRESS takes its name from Robert Burns, whose little collie Luath (*Gael.*, swift or nimble) tripped up Jean Armour at a wedding and gave him the chance to speak to the woman who was to be his wife and the abiding love of his life. Burns called one of *The Twa Dogs* Luath after Cuchullin's hunting dog in *Ossian's Fingal*. Luath Press grew up in the heart of Burns country, and now resides a few steps up the road from Burns' first lodgings in Edinburgh's Royal Mile.

Luath offers you distinctive writing with a hint of unexpected pleasures.

Most UK and US bookshops either carry our books in stock or can order them for you. To order direct from us, please send a £sterling cheque, postal order, international money order or your credit card details (number, address of cardholder and expiry date) to us at the address below. Please add post and packing as follows: UK – £1.00 per delivery address; overseas surface mail – £2.50 per delivery address; overseas airmail – £3.50 for the first book to each delivery address, plus £1.00 for each additional book by airmail to the same address. If your order is a gift, we will happily enclose your card or message at no extra charge.

ILLUSTRATION: IAN KELLAS

Luath Press Limited
543/2 Castlehill
The Royal Mile
Edinburgh EH1 2ND
Scotland

Telephone: 0131 225 4326 (24 hours)
Fax: 0131 225 4324
email: gavin.macdougall@luath.co.uk
Website: www.luath.co.uk